Commercial Vehicle Technology

Series Editor
Michael Hilgers, Weinstadt, Baden-Württemberg, Germany

Michael Hilgers

Electrical Systems and Mechatronics

Second Edition

Michael Hilgers
Daimler Truck
Stuttgart, Germany

ISSN 2747-4046   ISSN 2747-4054   (electronic)
Commercial Vehicle Technology
ISBN 978-3-662-66717-0   ISBN 978-3-662-66718-7   (eBook)
https://doi.org/10.1007/978-3-662-66718-7

This Springer Vieweg imprint is published by the registered company Springer-Verlag GmbH, DE, part of
Springer Nature.
The registered company address is: Heidelberger Platz 3, 14197 Berlin, Germany

Preface

For my children Paul, David and Julia and my wife Simone,
who share my passion for trucks.

I have worked in the commercial vehicle industry for many years. Time and again I am asked, "So you work on the development of trucks?" Or words to that effect. "That's a young boy's dream!"

Indeed it is!

Inspired by this enthusiasm, I have tried to learn as much as I possibly could about what goes into making commercial vehicles. During my tenure, I have discovered that you have not really grasped the subject matter until you can explain it cogently. Or to put it more succinctly. "In order to really learn, you must teach." Accordingly, as time went on I began to write down as many technical aspects of commercial vehicle technology as I could in my own words. I very quickly realized that the entire project needed to be organized logically, and once that was in place the basic framework of this series of booklets on commercial vehicle technology practically compiled itself.

This booklet deals with the mechatronics in the truck. The term mechatronics is explained comprehensively in the text. The short version in a nutshell: this booklet essentially describes the truck's electronic systems and brakes. Electric drivetrains are not considered here but are covered in one of the other booklets. Mechatronics in automotive vehicles is always a fascinating topic, and even more so at the moment with more advanced driving systems and automated driving are being developed! So, this second edition of the booklet mechatronics covers this area of the mechatronic system more comprehensively than the first edition.

Readers who are studying mechatronics for commercial vehicles (students and technicians) will find this booklet to be a good entry point and as a result may discover that commercial vehicle technology is a fascinating field of work. In addition, I am convinced that this booklet will provide added value for technical specialists from related disciplines who would like to see the big picture and are looking for a compact and easy-to-understand summary of the subjects in question.

My most important objective is to familiarize the reader with the fascination of truck technology and make it enjoyable to read. With this in mind, I hope that you, dear readers, have much pleasure reading, skimming and browsing this booklet.

Finally, I have a personal favor to ask. It is important to me that this work continue to expand and mature. Dear reader, I would greatly welcome your help in this regard. Please send any technical comments and suggestions to the following email address: hilgers.michael@web.de. The more specific your comments are, the easier it will be for me to understand its implications, and possibly incorporate them in future editions.

My wish is that everything is comprehensible and engaging. Happy reading!

November 2022   Michael Hilgers
                                                    Weinstadt-Beutelsbach,
                                                          Beijing, Aachen

Contents

Electrical Systems and Mechatronics

1

Even the motor vehicle patented by Carl Benz in 1886 already included an electrical component: the battery with make-and-break ignition. The radio could be considered one of the first complex electrical systems, which was installed in US vehicles around 1940. The first true electronic systems with complex semiconductor circuits were implemented in engine controllers in the 1960s. Since then, the electrical and electronic systems in motor vehicles have become progressively more important. Many of the vehicle functions that are standard today are only possible with the aid of sophisticated electronics. It is estimated that about 90% of all innovations in motor vehicles rely on electronic systems [1]. Figure 1.1 shows how the number of electronic control units (ECUs) in trucks has increased in the period between the late 1980s and 2014. This indicates the increasing importance of electronics in motor vehicles.

The importance of electronics will continue to grow in the future, but the absolute number of ECUs probably won't. Due to the constraints of space requirements, complicated wiring, etc., the number of actual ECUs will not increase at all, or will only increase very slowly. Instead, the performance capability of each individual ECU will continue to increase, as it has done in the past. In addition, ECU functionality will be devolved to the sensors and actuators, which are already referred to widely as intelligent sensors and actuators.

In the following sections, first the electrical infrastructure of the vehicle will be explained; then the mechatronic systems in a commercial vehicle will be described. Many components of the vehicle that are typically discussed under other headings can now quite legitimately be understood as mechatronic systems. For example, strictly speaking the hybrid, the start-stop function, the automated gearbox and even the engine itself are by definition mechatronic systems. But here we will apply the term as it is understood in everyday language in which the term mechatronics serves as a catch-all for the vehicle's electronic data processing (EDP) infrastructure, the brake system, electronic brake and chassis systems, advanced driver assistance systems (ADAS) up to

© Springer-Verlag GmbH Germany, part of Springer Nature 2023
M. Hilgers, *Electrical Systems and Mechatronics*, Commercial Vehicle Technology,
https://doi.org/10.1007/978-3-662-66718-7_1

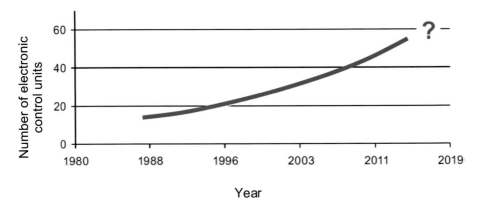

Fig. 1.1 Number of electronic control units (ECUs) in a European long-distance haulage truck in the last 25 years

autonomous driving and infotainment. The electric drivetrain, which by definition qualifies as a mechatronic system as well, is not included into the narrow definition of mechatronics used here. For the electric drivetrain see other books e.g. [4].

1.1 Wiring Harness

There will be many meters and several kilos of wiring installed in any given truck. There are wires used for supplying the necessary electrical power to the ECUs, sensors and actuators, and cables that serve to transmit signals. So a distinction can be made between the energy network (or power network, or powernet) and the communication network. Some sporadic attempts have been made to provide power supply and data over the same wire, similar to the solutions for building technology. However, the separation of communication and power supply is established as the more practicable approach.

The wires inside a harness connect the units of the mechatronic system—that is to say the ECUs, the actuators and the sensors. The interface between the harness and the mechatronic component typically has the form of plug-in connectors. Standardized connectors do exist; there are also connectors that have been developed specifically for a given connection. The plug-in connector provides a connection that can easily be made and disconnected again. This is important when it comes to maintenance (parts replacement). The plug-in connector must be designed in such a way that it transmits the signals and/or the electrical power reliably and does not become disconnected unintentionally (i.e. loss of contact-type problems). Depending on their installation location, plugs and plug-in connectors must also be able to withstand high temperatures, aggressive medias, dirt, moisture and vibration.

Given the enormous number of variants, equipment options and vehicle lengths, essentially every vehicle needs a unique wiring harness. In automotive production, one of two approaches may be adopted to deal with this diversity: In the first variant, harnesses are indeed assembled and installed in the vehicles specifically in association with a given purchase order. This means the harness only fits in the one specific vehicle for which it was made. The other methodology consists in defining several different standard harnesses. The vehicle is then fitted with the standard harness which is just sufficient to ensure the supply to the components in the vehicle. Accordingly, some vehicles then receive a harness in which wires are provided but do not serve in function in that particular vehicle. Thus, the fitting of unused wires is seen as worthwhile in order to reduce the number of variants of the wiring harness.

1.2 Power Supply

The power supply and distribution in medium-duty and heavy-duty trucks is a nominal 24 V DC voltage network in Europe and 12 V DC in the USA (more likely 28 and 14 V in real life). The voltage of the power supply is selected so as to ensure that it does not represent an electrical hazard for people. Due to the growing need for more electrical energy in the vehicle, the question is raised time and again as to whether the voltage should be increased to 42 or 48 V, for example, and there is justification for this. But as all components are available in the current voltage range, higher voltage values have not been established yet in the automotive world.

Some circuits in the onboard electrical system are defined in DIN 72552 [5]. For example, the permanent supply of voltage from the battery (i.e. positive terminal) is referred to as battery power or terminal 30. When the battery voltage is supplied as soon as the ignition is switched on, it is referred to as ignition power or terminal 15. Terminal 50 or crank power supplies voltage to the starter motor.

Electrical power for the onboard electrical system is delivered by the alternator or the (low voltage[1]) battery. The alternator is driven mechanically by the internal combustion engine, so it only supplies energy while the internal combustion engine is running. When the engine is not running, the battery is the only energy source. The alternator generates AC voltage, which is rectified.

When the ignition is switched on, normally all electrical functions of the vehicle are available. The engine does not necessarily have to be running. The electronics system is awake. Depending on how it is equipped, the total battery current of a vehicle with 24 V control system when it is awake is around 3 A. These 3 A are required simply to ensure

---

[1]The battery electric truck as well as a fuel cell truck have a high voltage battery that supplies electric energy for the drive. Still a low voltage 24 V battery is foreseen for the electrical control system of the truck (see [4]).

Table 1.1 Power consumption by various consumers that also run when the vehicle is stationary and place a load on the battery

	Output in watts (W) (approx.)	Power consumption in amps (A) in a 12 V network (approx.)	Power consumption in amps (A) in a 24 V network (approx.)
Vehicle awake, full equipment		≈6 A	≈3 A
On-board electronics asleep		≈0.2 A	≈0.1 A
Low beams, halogen (one side)	75 W	≈6 A	≈3 A
Reading light	10 W	≈0.8 A	≈0.4 A
Storage compartment light	5 W	≈0.4 A	≈0.2 A
Windshield heater	1000 W	≈85 A	≈42 A
Radio	30 W	≈2.5 A	≈1.25 A
Refrigerator box (average) depending on insulation and outside temperature	50 W	≈4 A	≈2 A
Coffee machine	300 W	≈24 A	≈12 A
Microwave	750 W	≈60 A	≈30 A

that the ECUs are ready to communicate and function. If additional electrical functions on the vehicle are used, such as the radio or the lights, or if accessories are powered via the sockets (coffee machine or the like), power consumption increases correspondingly. The power consumption and the current of certain consumers that are typically operated when the vehicle is stationary (as well) are shown in Table 1.1.

If the engine is not running, power must be supplied by the battery. The total usable charge that can be drawn from a battery is specified in amp-hours Ah.[2]

This charge is often also called the battery capacity. Typical battery capacities for commercial vehicles are in the order of 220 Ah for vehicles with high power requirements. Since a 220 Ah battery weighs over 50 kg (approx. 110 lbs), smaller trucks, trucks with minimal equipment and trucks used for particularly weight-sensitive tasks are equipped with batteries that have a lower capacity (100 Ah, 140 Ah or 170 Ah).

---

[2] For readers who are not well-versed in electrical terms: If a current of one amp (1 A) flows for an hour (1 h), this means that a charge of one amp-hour (1 Ah) is being drawn from the battery.

If 2 A flow for 3 h, this means that 6 Ah have been drawn. Current in [Amperes] is defined as charge per unit time. Thus, current times time equals charge.

With a power consumption of 6 A (in the 12 V case) or 3 A (24 V system), this means that just in the awake state the vehicle electronics will drain a fully charged battery in a matter of a few days.[3] To prevent this, the ECUs restrict their own activities when the ignition is switched off: The vehicle goes into sleep mode. This organized powering down and up of the ECUs is called network management.

Commercial vehicles in North America have a substantial challenge in starting heavy duty diesel engines in cold environments due to 12 V energy sources compared to 24 V in other parts of the world. The size of the engine is a big factor in determining how many and what size batteries are required. This often results in two to four 12 V batteries in parallel to meet cold cranking amp rating (CCA).

1.2.1   Network Management

When the vehicle is switched off (or ignition off), the electronics system attempts to power down the electrical and electronic systems as quickly as possible to save power and not drain the battery unnecessarily. However, the mechatronic system doesn't shut down immediately. The electronics terminate any processes that are still running and certain data is saved in the ECUs. Only when all ECUs report that they are not waiting to send or receive data and all internal processes have been terminated, will the ECUs shut down. The electronic logic that orchestrates all this is called network management. Ultimately, the effect is that the vehicle electronics do not go to sleep until some time after the ignition has been switched off.

Quiescent current

Even after the vehicle ignition has been switched off, the vehicle's electronic system must still be able to execute certain functions. For example, the onboard receiver for the remote control key must still be responsive so that the vehicle can be unlocked or locked. The alarm system, the clock and the tachograph need electrical energy. The alarm system (and the locking function) must also be able to actuate the vehicle's horn and light systems. This is why many ECUs are wake-up capable, meaning that they can be reactivated even when the ignition is switched off. For this purpose, the bus receivers of these ECUs must respond to a wake-up signal even when the ignition is switched off. These wakable ECUs need to be able to draw a small amount of power in the quiescent state, when the vehicle is deactivated as well, corresponding to a current of about 200 µA per unit. The whole vehicle needs a quiescent current (or standby current), roughly on the order of about 50 mA for a fully equipped vehicle.

---

[3] If more consumers are operated, the battery will be drained that much faster.

1.2.2   Starter Battery

The energy for starting the internal combustion engine is normally supplied by a device called the starter battery, or simply known as the battery. The battery for a large internal combustion engine needs very high currents for a short time. The lead accumulator technology typically used for the battery excels in delivering high current and power for a short period of time.

The battery supplies the consumers in the vehicle with electrical power even when the alternator does not, for example, when the engine is not running.

And besides its electrical requirements, the battery must also fulfill the basic requirements of the truck. It must be able to perform its duties at $-30\,°C$ and $70\,°C$ and withstand the vibrations acting on any component that is mounted on the vehicle frame.

Starter batteries are also called secondary batteries, which means they can be recharged, unlike primary cells, which cannot be charged a second time.

When the internal combustion engine is running, the battery is recharged by the alternator. The starter battery must be able to continue working after many such recharging and discharging cycles; this is called cycle strength. The charging process works well or not so well, depending on the temperature. With conventional lead-acid batteries, charging takes place very slowly at low temperatures. Unfortunately, the battery is subjected to greater loads in winter and is repeatedly recharged much more slowly than in warm weather. This explains why battery problems occur more frequently in winter.

During charging, the electrical energy that is fed to the battery is converted into chemical energy. In the inverse chemical reaction the stored energy is released again as electrical energy. There are many material pair combinations that are currently used in batteries or being researched for future use. A large number of material pairs are being evaluated as part of the discussion surrounding electric and hybrid vehicles [4].

As to the battery, the lead accumulator has represented the de facto standard for decades. It is fairly tolerant with regard to handling, it can withstand the thermal stresses inside the vehicle (i.e. does not need active cooling) and it is relatively inexpensive to produce. The lead acid battery is capable of delivering the high currents needed for the starting operation. The lead-acid battery can be recycled very well. In many countries, the recycling rate of the starter battery is very high, so that the ecological footprint of the starter battery is acceptable—even though lead itself is an environmental toxin.

The chemical reaction takes place between lead, lead oxide and sulfuric acid.

The reaction for delivering electric current (discharging) takes place as follows. At the battery's negative terminal, which consists of a lead plate, the lead is oxidized by the sulfuric acid, forming lead sulfate (Eq. 1.1):

$$Pb + H_2SO_4 \rightarrow PbSO_4 + 2e^- + 2H^+ \qquad (1.1)$$

At the same time at the positive terminal—a lead oxide plate—the lead oxide is reduced to form lead sulfate (Eq. 1.2):

$$PbO_2 + H_2SO_4 + 2e^- \rightarrow PbSO_4 + 2O^{2-} + 2H^+ \tag{1.2}$$

The total reaction can be formed from the sum of Eqs. 1.1 and 1.2. Equation 1.3 results:

$$Pb + PbO_2 + 2H_2SO_4 \rightarrow 2PbSO_4 + 2H_2O + \text{electrical energy} \tag{1.3}$$

Accordingly, in a lead accumulators' cells lead and lead oxide are transformed into lead sulfate. And the sulfuric acid is consumed. As a result, a fully charged starter battery contains a high concentration of sulfuric acid, whereas the discharged lead battery has a low concentration of sulfuric acid. The sulfuric acid concentration in a fully charged lead battery is in the order of about 35%. In these conditions, the density of the sulfuric acid is about 1.28 g/cm^3. A drained lead battery has a sulfuric acid density of about 1.1 g/cm^3. Thus the density of the acid is an indicator for the charge level of the battery.

During the charging operation, the reactions take place in the opposite direction under the effect of an external voltage. The chemical processes in an accumulator cell are illustrated in Fig. 1.2.

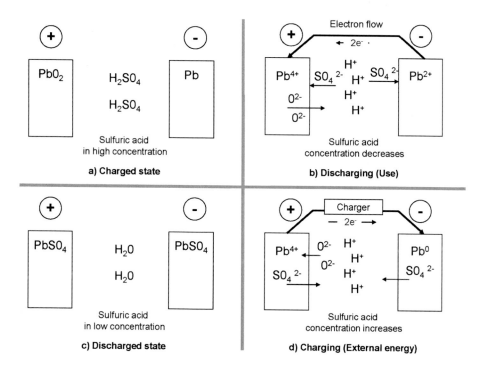

Fig. 1.2 Chemistry of the galvanic cell in a lead accumulator. **a** Charged state, **b** Discharging (use), **c** Discharged state, **d** Charging (external energy)

The difference in electric potential between the positive and negative terminals in a lead-sulfuric acid cell is a little more than 2 V. It follows that in order to create a battery with a rated voltage of 12 V, six lead accumulator cells have to be connected in series. If an onboard voltage of 24 V is required, this voltage is generated with two batteries of 12 V each connected in series.

As was mentioned earlier, trucks are equipped with batteries with a capacity from 100 to 220 Ah. The total charge that can really be drawn from the battery depends on the temperature, the aging of the battery and the discharge conditions. Low temperatures have a particularly strong tendency to reduce the usable capacity and the maximum available current of the battery.

The larger the volume of lead and lead oxide, the greater available charge and therewith also the usable energy content of the battery.

The maximum theoretically available charge of a lead accumulator per unit of weight can be estimated as follows. The reagents on the left side of the chemical Eq. 1.3 have the following molar masses—Pb: 207 g/mol, PbO_2: 239 g/mol and H_2SO_4: 98 g/mol. But since the sulfuric acid is a 33% solution, it must be included in the calculation at a level of about 300 g/mol per molecule. Since two H_2SO_4 molecules are needed, one mole of the starting materials weighs $207\,g + 239\,g + 2 \cdot 300\,g \approx 1000\,g = 1$ kg. Given that two electrons flow per reaction, one mole or one kilogram of accumulator mass delivers the following maximum charge[4]:

$$2 \cdot 6.022 \cdot 10^{23} \cdot 1.6022 \cdot 10^{-19}\,C = 1.93 \cdot 10^{5}\,As = 54\,Ah \qquad (1.4)$$

The maximum theoretical charge quantity of a lead accumulator is in the order of about 54 Ah per kilogram mass. Since six cells are connected in series in a 12 V battery, the available charge is reduced to about 9 Ah per kilogram. Modern lead accumulators typically provide about 4 Ah per kilogram.

Given that the cell of the lead accumulator delivers about 2 V, the theoretically available energy density in the lead accumulator storage medium is roughly 110 Wh per kilogram:

$$2\,V \cdot 54\,Ah = 108\,VAh = 108\,Wh \approx 0.4\,MJ \qquad (1.5)$$

The actual gravimetric energy density is still significantly less than 0.4 MJ/kg, because passive components and the housing add to the overall weight. Values for other energy storage units are discussed in [4]. There it is shown that the lead accumulator is not suitable for use as a storage unit for driving energy.

The battery ages with use. An aged battery no longer has the same rated capacity as a new battery. This is why the quantity called state of health (SOH) has been defined

---

[4] 1 mol consists of $6.022 \cdot 10^{23}$ particles and the elementary charge is $1.6022 \cdot 10^{-19}$ C.

to give a measurement of the battery's aging condition. The SOH is derived as a ratio between the maximum energy storage content still possible and the rated capacity:

$$SOH = \frac{\text{Currently possible maximum capacity}}{\text{Rated capacity}} = \frac{C_{max}}{C_{Rated}} \qquad (1.6)$$

The charging state of a battery is often referred to as the state of charge (SOC), it is calculated as a ratio between the charge in the battery and the maximum capacity of the battery possible currently after aging:

$$SOC = \frac{\text{Current state of charge}}{\text{Current maximum possible capacity}} = \frac{C_{current}}{C_{full}} \qquad (1.7)$$

In order to be able to determine the condition of the battery, battery sensors are used. The battery voltage and the ambient temperature are measured, and the currents flowing out of or into the battery are determined continuously. The condition of the battery is deduced from these variables with a simulation model. By its nature, modeling the condition of a battery is associated with some degree of uncertainty.

1.2.2.1 Battery Cable

The battery cable must conduct very high currents when the engine is started. The starter current for heavy trucks with 12 V network may be more than 1200 A for several seconds, briefly even higher than 2000 A. For this reason, a low-resistance cable (that means thick cable) is used. These cables are traditionally made of copper. Because the cable is so thick, the volume of metal in the cable is noticeable. Aluminum is lighter and the price of the raw materials is lower, so in the future it can therefor be expected that the use of aluminum in the starter cables will become more popular.

1.3   Lighting

Lighting and signaling systems are mandated by law and are required on every vehicle. Figure 1.3 lists a range of different lighting functions of the vehicle's external light system.

Low beam headlamps are used to illuminate the road for the driver and to make the vehicle more readily visible to other road users. The light cone of the low beam headlamp is designed asymmetrically so that the outer edge of the road is lit for a considerable distance while the middle of the road is not illuminated so strongly to avoid blinding oncoming traffic.

High beam headlamps illuminate the road for a much greater distance in front of the vehicle than the low beams, but they must be deactivated for oncoming traffic because there is a risk of blinding drivers coming towards the vehicle. When the high beams are switched on, a blue indicator lamp lights up on the instrument panel, also known as the dashboard.

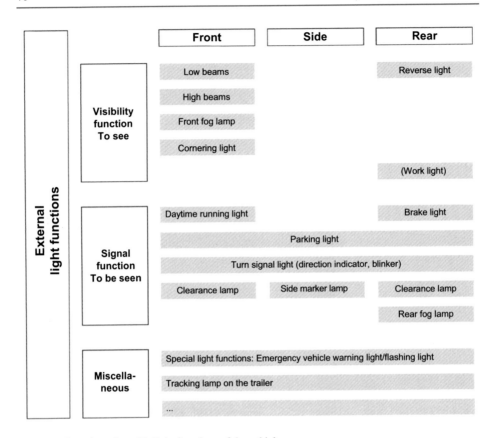

Fig. 1.3 Overview of outside light functions of the vehicle

The main headlamp includes both the low beam and the high beam lights in most modern cars. Other lights such as fog lamps or blinkers can be integrated in the housing.

In order to create the light cone using the low beam and high beam lights, two different technologies are employed. First, the light source (the lamp) is mounted in front of a reflector, which determines light intensity and the shape of the light cone. Secondly, some lamps also use a lens as well as the reflector to create the best possible light cone.

The headlamp cover also fulfills a lighting function. Structured regions of the cover prevent glare and scatter the light to produce a more pleasant, softer lighting profile.

Light sources used in the main headlamp are H4 lamps, H7 lamps or Xenon lamps. H4 and H7 lamps are halogen lamps, the H7 is a newer, stronger version of the halogen lamp. The light flux from xenon lamps is more powerful still. Xenon lamps are currently offered as special equipment.

In vehicles in the passenger car segment, more and more LED headlamps[5] are being used. Unlike the xenon lamp, the LED does not need a warm-up phase. LEDs also allow

---

[5] Light-emitting diode, LED.

very pleasant spectral light distributions. LED lamps require a stable operating voltage, otherwise they age prematurely. LED solutions are still expensive at the moment.

The light yield from LED lamps (conversion efficiency from electrial energy to visible light) is much better than with conventional bulbs, so the total quantity of heat released is lower. However, the heat given off is concentrated in a very small area, which means that heat sinks and even active cooling with fans are used to cool the thermally heavily loaded LEDs. The low quantity of total heat output by the LED lamps creates an additional problem in winter. Iced LED headlamps are difficult to defrost.

(Dim) LED lamps are widely used as side marker lamps on trailers. LED technology is also offered for daytime running lights for commercial vehicles.

Daytime running lights are not as bright as the low beam headlamps and are used for signaling. Daytime running lights help to make vehicles visible earlier and clearer, thus reducing the risk of accidents. In some countries daytime running light is mandatory for trucks.

1.3.1 Monitoring the Lighting Function

A system for monitoring the function of blinkers (or turn signal indicators) is required by law. Light source monitoring for other lighting units (i.e. low beams, high beams, tail lights) is also offered in a number of vehicle models on the market today. Typically they are designed to detect three faults: an interruption in the conductor track (or open load) is one sign of a faulty light source, short circuit to ground, or short circuit to the battery voltage are indicators of other faults.

1.3.2 Automated Light Functions

In order to improve visibility while driving, an active lighting system can be used. The **cornering light** illuminates in the direction in which the vehicle is being steered when cornering. Thus, the relevant section of road has improved visibility.

The **automatic high beam** function uses a camera system to determine how far ahead the road should be illuminated and adjusts the headlamp reflector accordingly. With oncoming traffic or on well-lit roads, the automatic high beams are deactivated.

1.3.3 Interior Illumination

When driving, the interior illumination is designed to help the driver's orientation and ability to make out the displays and control elements easily. The instrument panel is illuminated, and switches and levers must also be clearly visible in darkness. The switches are typically backlit. At the same time, the interior illumination must not blind or distract the driver while driving.

On the other hand, when the vehicle is stationary the interior must be well enough lit so that the driver is able to live in it. Bright interior illumination enables the occupant(s) to work and conduct everyday activities inside the cab. Ambient lighting is provided to create a comfortable and relaxing atmosphere. Special reading lights make reading easier for the occupant(s). Other interior lights are often located in the storage compartments and the footwell of the cab. Entrance lighting provides assistance when climbing into and out of the cab.

Various lighting functions are coupled with other functions. Some examples include the interior light that comes on when the door is opened, the low beam headlamp that lights up when the vehicle is locked or unlocked (a feature that illuminates the front of the vehicle for a prescribed time, the Follow Me Home function), or the turn signals flash when the vehicle is locked or unlocked with the keyless remote.

1.4   Horn and Other Acoustic Signals

The horn is a system required by law. The regulations for acoustic signals are set forth in ECE R-28. The simple electromagnetically-actuated horn functions according to the Wagner's hammer principle: When a current is applied to the electromagnetic horn it passes through a coil. A lever is attracted by the coil which causes the current flow to be interrupted again. The coil's magnetic field collapses and the lever returns to its resting position due to the restoring force of the disc spring. Now the current can flow again. The activation and deactivation of the coil and the spring force of the disc spring set up an oscillation which in turn causes the entire horn body to vibrate. The stiffness and rigidity of the disc spring and the rest of the horn body (including the bracket) are what determine the horn's tone.

Besides the electric horn—or instead of it—many trucks are also equipped with air horns. In air horns, compressed air is forced through a diaphragm or reed, causing it to vibrate. This produces a deep, very loud sound. If an electric horn and an air horn are installed at the same time, the driver can choose via a switch in the cab whether pressing the horn (typically on the steering wheel ring) actuates the electric or the pneumatic horn.

Air horns are often mounted on the roof in the form of chrome-plated ornamental features for an impressive appearance. Alternatively, compact horns operated by compressed air also exist. These can be fitted in the wheel arch, for example, so they are not visible from the outside. They have a similar deep sound to the roof-mounted horns. But the roof-mounted horns are very popular with customers because of their appearance. Customers are willing to pay the price of increased air resistance and greater fuel consumption for the visual effect of roof-mounted horns.

1.5   Building Blocks of Mechatronic Systems

Mechatronic systems include mechanical components, electronic components and an information processing unit. In essence, a mechatronic system consists of sensors which capture (or measure) a physical reality and convert it into electrical signals. These signals are processed in electronic control units (ECUs) and compared with specified values. The various ECUs exchange information among themselves; they are interconnected. The ECU can then transmit control signals to an actuator to trigger an action that affects a change in the mechanical world. This change is in turn detected by the sensors again. The basic principle of the mechatronic system is summarized in Fig. 1.4.

All components of the mechatronic system must be able to withstand the same harsh conditions of the commercial vehicle environment to which all truck components are exposed. These include extreme temperature fluctuations, moisture, possible aggressive media such as oil, salt water or diesel (i.e. fuel tank fill level sensor) and mechanical stresses due to vibration, for example. And beside these, there are other requirements that apply only to mechatronic components, for example electromagnetic compatibility—see Sect. 1.6.4.

The **sensors** measure a given physical quantity and output an electronic signal. Sensors exist for an enormous range of physical measures, including temperature, acceleration, engine speed, atmospheric humidity, brightness, pressure, and many more. In the sense of Fig. 1.4, the driver's operating elements are also sensors. Pressure on the gas

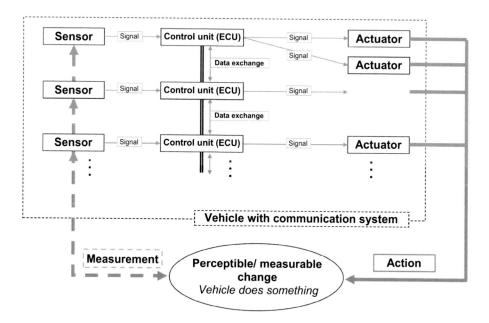

Fig. 1.4 Basic principle of a mechatronic system

pedal or operating a switch signals a driver input. The position or the change in position of an operating element is processed in the ECUs as a measurement value and prompts an action by the actuator.

The mechanical action of the system is executed by the **actuators**. Put simply, an actuator is a component that receives an electrical signal and returns a mechanical reaction or a visibly perceptible signal. Typical examples of actuators are electric motors, electrically powered hydraulic units, electromagnets, solenoid valves, piezo actuators, or even displays, loudspeakers and lights.

Nowadays, technically advanced commercial vehicles have over 30 **electronic control units (ECUs)** for processing the information from sensors and for controlling actuators. The complexity of the different ECUs varies considerably.

The large number of sensors, actuators and ECUs means that a highly complex system architecture and a very sophisticated **communication system** are indispensable. The ECUs normally communicate with each other via bus systems. In all cases, the exchange of information among ECUs is carried out digitally. Analog data from a sensor is digitized either directly in the sensor or in the ECU and processed and forwarded in digital form. Different levels of the mechatronic system are illustrated in Fig. 1.5.

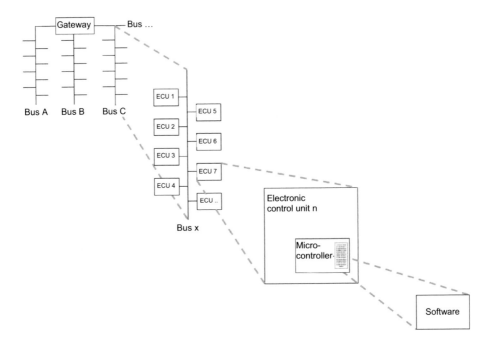

Fig. 1.5 Different levels of the mechatronics system in the motor vehicle

1.5.1 Data Bus Systems

A data bus is a shared data link between multiple devices designed to exchange data. Since multiple (many) devices exchange data via a common bus, the wiring arrangement can be kept relatively simple compared with point-to-point data links. Bus nodes can send data to the bus and receive data from it. Nodes that only receive data are passive nodes. Nodes that send as well are active.

The amount of data that is exchanged in a well-equipped modern motor vehicle is so big that a single bus for the entire vehicle would be overloaded; so, a vehicle is usually fitted with several bus systems (sub-buses). It makes sense to arrange electronic control units (ECUs) on the same bus that exchange information frequently. Data is exchanged between the different bus systems via gateways. Because of the open construction of truck frames, it is inevitable that some data bus lines are accessible from the outside. Here too, it is helpful to share the data exchange tasks among several different sub-buses: ECUs for which it is particularly important to provide protection against outside access, such as the locking system ECU, are connected to bus lines that do not leave the driver's cab.

There are several different bus technologies that will be discussed below. A whole range of different bus technologies is implemented for the various sub-buses in a vehicle. The technology for a given sub-bus is selected on the basis of various requirements, such as availability, cost and data rate capability.

The practice of installing bus systems has been widespread throughout the automotive industry for a long time. In principle, the same systems are fitted in trucks as are used in much larger numbers in the passenger vehicle industry. But there are also some requirements exclusive to bus systems that are to be installed in trucks. For example, the bus line in a truck can be very long. When designing a bus system for a truck, one must bear in mind that product service life in the truck industry is very long; vehicle models will be produced for more than 15 years. It must be possible to expand or upgrade the bus system to keep pace with the development of the truck so that it will still be compliant with the (predicted) requirements applicable to a bus system in 15 years.

An important design criterion for a bus system is the maximum volume of data that can be transmitted per unit of time, also called the baud rate, or more commonly the "bandwidth". The bandwidth must be sufficient to handle the volume of data that is to be transmitted via a bus. Transmitted data volume divided by bandwith gives the bus load. This depends not only on the number of ECUs but also on the repetition rate at which information is placed on the bus (cycle time) and the resolution at which the information is sent. Experience has shown that the amount of information that has to be transmitted on a bus increases steadily year on year, as new functions and ECUs are added. Experienced developers expect the bus load to rise by about 3% per year on average.

Not all the information supplied by the ECUs is equally urgent. So the specification as to when which information is put on the bus is very important. This is called bus access.

The various bus systems implement different access procedures, so bus access can be considered a decisive (but not the only!) characteristic of bus systems.

Random access procedures do not specify which ECU can put data on the bus at what point in time. All units can send at any time when another unit is not currently sending. A suitable procedure must be defined to assign a priority when two ECUs start sending at the same time. For systems with random access procedures, it is not clearly defined at what time or with what delay an information is transmitted to other ECUs.

The most important bus system currently in use, the controller area network (CAN) bus works on the random access principle.

Systems with controlled access procedure work differently. In controlled systems, it is specified when a given ECU is permitted to send data. This ensures that an item of information will be sent via the bus after a defined period. The local interconnect network (LIN) and FlexRay are two bus systems with controlled access.

In controlled bus systems, access control can be centralized or decentralized. With centralized control, one electronic control unit—the master—has a higher level function and instructs the other units to transmit their information. If control is decentralized, a communication sequence is defined and all ECUs obey this sequence. The obvious advantage of centralized control is that if the system is changed or expanded (new ECU = new node on the bus) only the master has to be changed. If changes are made to systems that are decentralized but controlled, all of the ECUs in the system must be adapted.

Besides the logical differences between the bus systems, there are also considerable physical differences. For example, the bus systems installed in motor vehicles vary considerably in terms of signal voltages and data transmission rates.

Table 1.2 lists a few widespread bus systems. References [1, 2] contain further explanaitions of several bus systems. The CAN bus and the LIN bus are currently both widely used bus systems in commercial vehicle technology.

Table 1.2 Various bus systems commonly used in commercial vehicles, and possible buses of the future

	Data rate	Comment
LIN	20 kbit/s	
Lowspeed-CAN	125 kbit/s	
Highspeed-CAN	Up to 1 Mbit/s	Line lengths typical for trucks
FlexRay	Up to 10 Mbit/s	
Classic Ethernet	10 Mbit/s	
Fast Ethernet	100 Mbit/s	
Giga-Ethernet	1000 Mbit/s	

1.5.1.1 LIN Bus

The LIN bus [13] is a commonly used standard for simple applications that do not involve transmitting large data volumes. In particular, actuators, sensors and switches are often linked via a LIN bus.

The LIN bus functions with centrally controlled communication. The network consists of a master and one or more slaves. The master specifies which slave is permitted to send data via the bus. To do this, the master sends a header addressed to the slave requested. The slave sends the data in its response. If additional slaves are integrated in the system, relatively few changes have to be made. Only the master needs to be adapted; for the slaves, nothing changes.

The master contains a table (the schedule) according to which the master queries the various slaves. This serves to define the load on the bus and the repetition rate of a signal is guaranteed. Collisions on the bus cannot happen. The master may include several schedules, and it can switch between them according to the situation.

Each bus node (master or slave) is able to decide whether it processes the message or not. This is a system with selection by the recipient. Accordingly, a message can be accepted by one, several or even all bus nodes.

The maximum bandwidth for a LIN bus is 20 kbits per second.

LIN systems can be configured at quite low cost, so they are used very often.

1.5.1.2 CAN Bus

For more complex data exchange, the CAN bus (see [10]) has established itself as the dominant bus system in the last twenty years. There are two varieties of the CAN bus[6] in use, the high-speed CAN and the low-speed CAN. The low-speed CAN offers a lower data rate, but its structure is extremely sturdy and fault tolerant.

The CAN bus works with random access procedures. If several ECUs are attempting to place data on the bus, an arbitration procedure specifies which unit has the highest priority.

Arbitration

The arbitration procedure is based on the fact that there are dominant and recessive bits. Each ECU attempting to place a message on the bus begins with an identifier, a sequence of bits. If two or more nodes are attempting to send messages to the bus, they both send their respective identifiers at the same time. If one (or more nodes) has/have a dominant bit, this signal level prevails on the bus.

By definition, the identifier which begins with the longest sequence of dominant bits has the highest priority. If an ECU sends a recessive bit in the arbitration phase and at the same time detects that the signal level on the bus corresponds to a dominant bit, the unit recognizes that another ECU is sending a dominant bit and therefore has higher priority. The unit with the recessive bit withdraws from the arbitration phase.

---

[6] CAN = Controller Area Network.

1.5.1.3 Other Bus Systems

In order to increase the rate at which data can be transmitted but still keep the tried and tested CAN protocol, the newer CAN bus extension **CAN FD** is implemented [11, 12]. CAN FD stands for CAN with flexible data-rate. The arbitration phase remains the same. With CAN FD, the bits (after arbitration) are transmitted at a higher rate, and CAN FD has the capacity for longer data fields in a frame, so the ratio of data rate to bitrate (bitrate = data + overhead) is more favorable. Since CAN FD is uncomplicated in terms of both hardware and handling and retains the proven advantages of CAN, the CAN FD protocol can be implemented easily.

FlexRay has been in use as a bus system in cars for several years. The notable feature of Flexray bus systems is that the cycle time for an information item is guaranteed. The information will be transmitted in a given period of time without fail. FlexRay also supports two-channel information transmission and so it ensures redundant communications. Consequently, FlexRay systems can be upgraded (at additional cost) until they satisfy the safety requirements for by wire systems. FlexRay components available now permit higher data rates than the conventional CAN bus, although both are physically similar in terms of hardware. But it is also true that a FlexRay bus is considerably more elaborate to implement than a CAN bus system.

Ethernet is also a possible bus system for automotive applications. FlexRay already has a high data rate, but Ethernet is capable of rates in the order of roughly ten times better. Here too, the passenger car sector is leading the automobile industry. There are cars in which subnets are configured with Ethernet.

The **PSI 5** standard[7] [14] offers simple, inexpensive data transmission for integrating sensors, for example, and can be used for point-to-point data transmission or simple bus configurations. Only two lines are needed for supplying power to the sensor. The signal is modulated onto the power lines. The result is affordable data transmission. It is weaker than the expensive bus systems in terms of electromagnetic compatibility (EMC) and transmission rate.

The **I²C-Bus** is also very simple. Unlike the PSI 5, this system was not originally intended for use in motor vehicles. It was developed for home electronics [15].

1.5.2 Electronic Control Units

ECUs are the devices where the information is actually processed. The control programs of an ECU run on processors called microcontroller (often abbreviated to μC). Software is loaded onto the microcontrollers. A high-level language often used for programming the microcontrollers is C or C+. ECUs are being programmed more and more often using models. The function developer describes the desired function in a model and this is translated into executable program code by an autocode generator.

---

[7] Peripheral Sensor Interface 5.

1.5.3 Autosar

AUTOSAR (stands for AUTomotive Open System ARchitecture) [9] is an open standard for designing the architecture of vehicle electronics. The central idea is to design the ECU hardware and the applications that run on the ECU separately from each other. The intention is to enable software components to be moved without difficulty from one ECU to another, or equally to offer the possibility to change the ECU hardware without having to adapt the application software. This means that the software architecture in the ECU is modular with defined interfaces. Ideally, application software components in an AUTOSAR-compliant architecture can easily be removed from one ECU and plugged into another.[8] In this architecture, the application functionality is represented in the application layer independently of the hardware. Hardware-specific software scopes are realized in the basic software. A large number of basic software modules have been defined for this basic software in AUTOSAR. The Runtime Environment (RTE) supports the exchange of information between the various software components of the application layer and establishes the connection between the application layer and the software components specific for a given ECU (operating system etc.).

Figure 1.6 shows a highly simplified model of AUTOSAR. The layer of application software is decoupled from the hardware-dependent and hardware-associated basic software by the RTE. The basic software takes care of network services, bus communication, memory management, operating system, diagnostics functionality, etc.

The literature on AUTOSAR is diverse and abundant, as a visit to the *"News and Publication"* page on [9] will reveal.

1.5.4 Sensors

The number of sensors on a modern motor vehicle is large. A range of measurement principles is applied—see Tables 1.3 and 1.4. In principle, all electronic sensors ultimately measure a voltage. Voltage can be processed particularly well on electronic components. The sensor uses a suitable physical process to convert the measurement variable to a voltage.

The quality of a sensor signal is defined by the resolution of the sensor and the signal-to-noise ratio. For system developers, it is also important to know the magnitude of the sensor scatter, that is to say the magnitude of the difference between the output values from two identical sensors measuring the same input variable. The signals from the

---

[8] In practice, the developers must deal with some limitations due to the fact that resources for a given functionality and physical connectors (pins) must be provided in the ECU, and while some units have them, others do not. Signal runtimes and response times also tend to vary when software components are transferred from one control unit to another.

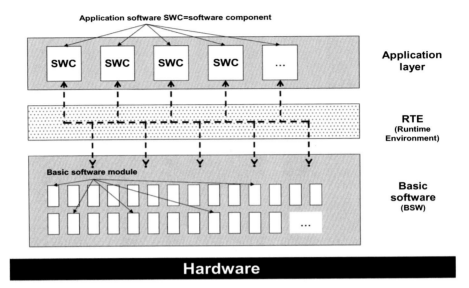

Fig. 1.6 Highly simplified layer model of AUTOSAR. More detailed illustrations of the architecture can be found in the pertinent literature and can also be obtained from the AUTOSAR consortium [9]

sensors are checked for plausibility and may be corrected if external influencing varia-bles such as temperature, etc., are known. Sensors with self-diagnostic function return a defined signal if the sensor is obviously measuring incorrectly.

Cameras, radar systems and lidars are sensors as well. Those complex sensors provide the input signals for imaging processes required by advanced driver assistance systems and are explained in the section on advanced driver systems—see Chap. 4.

1.5.5   Switches and Control Levers

Important inputs for the mechatronic system are the demands of the driver and the co-driver. These are input by means of foot controls, switches and push buttons, and con-trol levers. A distinction is made between load and signal switches. Load switches allow or interrupt the current flow directly. On the other hand, signal switches, when operated, send a signal either via a bus system or a special signal line to communicate the desired change in function to an ECU.

Table 1.3 Examples of different measurement principles and sensors in a truck (Part 1)

Measurement variable	Measurement principle	Example
Temperature	Voltage drop over a negative temperature coefficient (NTC) (temperature-dependent resistance)	External temperature sensor
Barometric pressure	Pressure-dependent resistance of a piezo sensor (piezo-resistive)	Input variable for engine control
Pressure up to 15 bar	Pressure-dependent resistance of a piezo sensor (piezo-resistive)	Tire pressure sensor
High pressure (= 400 bar)	A resistive bridge is seated on a metal diaphragm. If the shape of the metal diaphragm changes under pressure, the resistive bridge gets out of tune. Measurable voltage differences are generated	Pressure sensor in the fuel system
Angle and angular changes = rotational speed	Optical principle (contact free): LED emits light that is detected by light sensor (CCD)—coding ring between LED and sensor Hall principle: The rotating part is equipped with a sender wheel. Contactless scanning with Hall sensor. Reference point through discontinuity of hole pattern in the sender wheel	Steering angle sensor (needed for ESC) Rpm sensor (e.g. camshaft, wheel speed)
Acceleration	Change in capacitance of a micromechanical plate capacitor	Acceleration sensor of the airbag ECU
Angular velocity, yaw rate	A movable structure changes its position relative to another structure in response to the Coriolis effect Option A: The change in position is detected piezoelectrically Option B: The change in position is detected via the change in capacitance of a micromechanical plate capacitor	Sensor for ESP
Oxygen concentration (in exhaust gas)	Semiconductor material whose resistance changes according to the concentration of oxygen in the ambient gas	Lambda (oxygen) sensor for engine timing

Table 1.4 Examples of different measurement principles and sensors in a truck (Part 2)

Measurement variable	Measurement principle	Example
Air mass flow	The air mass to be measured flows over a heated sensor diaphragm. The temperature is measured at two measuring points. The difference in temperature is a measurement of the cooling air volume blowing by	Air mass measurement in the intake
Presence of water condensation	Heating and temperature measurement at two measuring points, thermal conductivity enables conclusions to be drawn about the surrounding medium	Condensation water sensor
Flow rate using oval wheel	The volume throughput of the liquid turns two oval wheel gears. The rotation of one gear is recorded by installed permanent magnet/metal (see Rotation measurement). The number of rotations is proportional to the flow	–
Liquid fill level with reed switch	A permanent magnet is attached to a float and guided past a chain of reed switches. The electronic analysis system detects which reed switches are closed	Fuel tank fill level sender
Fluid fill level by ultrasound	A piezo sensor emits an ultrasonic signal. This is reflected by the surface of the liquid and received again by the piezo sensor. The level of fluid is calculated from the transit time[a]	Measurement of fuel tank fill level
Urea concentration	In actual fact, it is not concentration of the liquid that is measured, but rather its density. A piezo-ultrasonic signal is directed at a reflector over a defined measurement distance. The transit time is proportional to the density of the fluid	Urea quality sensor

[a]The transit time is temperature-dependent, so it is logical to measure the temperature and do temperature correction on the measured time. If the measuring distance of the signal is not too short, the same piezo sensor can be used as both sender and receiver microphone. The sensor can be located in the fluid and send the signal upwards towards the fluid surface or it may be positioned in the air space above the fluid

1.5.6   Actuators

Actuators are the components that actually perform the physical changes requested by the ECUs. Typical actuators are electric motors and electromagnets. However, the answer to the question of what an actuator is depends on the context. With regards to a transmission controller that manages gear selection, the pneumatic valve in the transmission is the actuator. For the purposes of cruise control, the entire drivetrain is one complex actuator whose task is to set the desired travel speed.

1.6   The Overall Mechatronic System

In later sections, (beginning with Sect. 1.7), individual mechatronic systems are explained. However, the functionalities and properties of the vehicle mechatronics system in the system as a whole are discussed first.

1.6.1   Functional View

For the vehicle user, the individual components of the mechatronics system—the actuators, sensors and ECUs—are not important. The only thing that is important to the user is the function. The exceptions to this are the switches, control levers and the displays, because this is where the user is directly aware of the hardware component and decides whether they are good or not.

The specification of a mechatronic system when designing it and the description of the system explaining it to the customer follows the functional view rather than a component view. To do so one has to define "functions". The split up of the overall system in functions can be rather coarse or very detailed. Defining what is a function is not always unambiguously possible. There might be different options to split up the system in functions and subfunctions etc. As a good and easy to use approach the following hands-on definition proves to be helpful: A function is a capability of the system that has its own user interface (switch, lever, etc.) or has its own output (sound, light, display image) or has its own name that is used in communicating to the end customer. Many of the functions cannot be assigned uniquely to one component. These are called distributed functions. Examples of distributed functions are the immobilizer, which consists of a combination of the engine management, electronic ignition lock and transmission control. Another example is the emergency brake assistant. This involves coordinating the radar, a decision-making function of any kind which is built into an ECU, the instrument, the engine and transmission and the brakes. All contributing components need to work as one system. Even the drive function is a distributed function from a mechatronic point of view. It needs the engine, drive control, transmission control, electronic ignition lock, gas pedal, and other subsystems to work together to move the vehicle.

1.6.2 Architecture

The architecture of the electronics defines how the many mechatronic components work together. The functional architecture describes the functions that are included in the overall system without reference to their physical realization. It serves to define what information is required for the individual functions and what information each function supplies to the other functions. The result is a theoretical construct of functions and communication relationships.

A technical architecture is derived from the functional or logical architecture by assigning the functions or sub-functions to certain ECUs and actuators. The technical architecture defines the requisite data exchange pathways between the ECUs. It is logical and practical to assign related functions and functions that share large volumes of information to the same ECU.

The topology by which ECUs are networked is also defined at this point. Widely used topologies are the classic linear bus topology, the star topology and the ring topology. Figure 1.7 illustrates the various bus topologies. The overall vehicle usually contains a mixture of several basic topologies. The networking topology is also influenced to a degree by the bus technology chosen.

The technical architecture also includes the onboard electrical system with battery and cables that supplies the electrical energy to the sensors, ECUs and actuators.

1.6.3 Diagnostics and Flashing

Given the complexity and variety of causes for malfunctions or failures in electronic systems, it is imperative that the electronic system has a powerful diagnostics functionality.

For some applications, a diagnostics functionality is explicitly mandated by law; especially in the case of the exhaust aftertreatment systems, certain diagnostic functionalities are legally mandated.

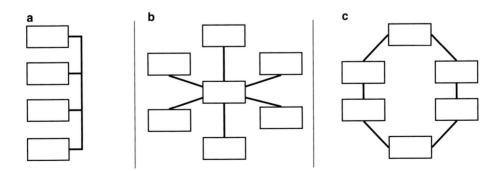

Fig. 1.7 Simple network topologies. **a** Linear topology, **b** Star, **c** Ring

In the generally accepted sense, a diagnostics function means that an ECU detects anomalies in the unit or at the inputs to the unit, evaluates them and stores them in an error memory together with a fault code[9] that describes the anomaly as detailed as possible. A fault code is stored in the ECU as a hexadecimal value. Selected fault codes are translated into understandable plain text—depending on the vehicle—and displayed for the driver on the vehicle's instrument panel.

The fault codes stored in the ECUs can be read out and interpreted via an external device, called a tester, which is plugged into a jack in the vehicle for this purpose. Modern testers can combine different fault codes from various ECUs to help analyze the underlying root cause of the problem. Some powerful devices also have their own menus that provide service technicians in the workshop with guidance as to what they need to test.

As more and more vehicles have a radio connection to the outside world, over the air diagnosis (OTA diagnosis) becomes increasingly important. Vehicle systems and functions can be checked and diagnosed remote over the air without having the vehicle in the workshop. The measurement values of the sensors connected to the ECU can be accessed via the diagnostic interface. The ECU can also be instructed to run certain routines such as self-tests via the diagnostics function.

The external access to the ECUs be it via a pluged-in tester or over the air is not only used to diagnose the system but can also be used to change the system by writing new parameters or new program code into one or several ECUs. The technique to replace or add program code or parameters to the ECU is called flashing. This is a way to update the ECU software and improve a vehicle function if necessary. The functionality of an ECU can be adapted to certain requirements, for example, if a vehicle is operated in another country where different regulations exist or in a different field of activity requiring the vehicle to behave differently.

If the flashing is done from remote via the radio connection it is called over the air flashing.

1.6.4   Electromagnetic Compatibility EMC

The propagation of electromagnetic fields can cause electrical and electronic devices and systems to interfere with each other.

In order to avoid these adverse influences, the electromagnetic compatibility (EMC) of a device must be assured. EMC is divided into two aspects:

---

[9] The term fault or fault code has become accepted over a period of decades, since the primary purpose of fault codes is to provide assistance for cause analysis and troubleshooting in the event of a malfunction. The fault code does not necessarily describe that something is broken. It might refer to abnormalities and events that provide assistance when analyzing the vehicle's behavior. A fault code "AdBlue tank empty" is helpful for analyzing the vehicle's behavior but does not mean that the vehicle needs repair.

- Interference emission—A device must not cause any impermissible influence on another apparatus.
- Interference immunity—A device must be invulnerable to external interference signals.

Requirements regarding the EMC properties of vehicle components are defined in e.g., the ECE regulations [8].

The influence of one device on another may be transferred via the wiring. In this case, the interference signal travels from one device to another through the electrical wiring; or it may be field-borne. An electromagnetic field is radiated by the interference source and a component of the affected device functions as an antenna.

Possible external sources of electromagnetic interference are powerful transmitters, for example. One natural source of interference is a lightning strike, which emits very powerful electromagnetic waves. But even relatively low-power devices such as those used in entertainment electronics can cause electromagnetic interference if they are not well designed and are moved close to a vehicle's electronic components.

A range of technical measures are possible for improving the interference immunity of a component:

- Suitable filter circuits increase the EMC immunity of electronic switching circuits.
- Wires and components can be shielded.
- The geometry of components and the wiring can be optimized with regard to EMC considerations. A simple example would be twisted wires.
- Signal transmission paths that work with differential signals are less sensitive to EMC interference.

1.7   Instruments and Displays

Besides the switches and control levers (Sect. 1.5.5), instruments and displays make up the interface between vehicle and operator. Together, this is referred to as human–machine interface (HMI). Whereas the switches and control levers are used by the driver as inputs, the task of the instruments and displays are to show the driver information as an output. Data on the vehicle operating conditions is presented. There is current driving information that the driver must have (i.e. vehicle speed, engine speed, mandatory lamps and turn signals), information about the condition of the vehicle (i.e. engine oil temperature, tire pressure monitoring, and fault codes), navigation aids and infotainment (i.e. radio stations and phone). Depending on the equipment included on the semitrailer or trailer and the vehicle, information about the trailer may also be displayed, for example, the pressure in the trailer tires or the temperature in a refrigerated body or trailer. The visual display with instruments and warning indicators is often supplemented with other signal paths: haptic (tactile) and acoustic signals support the visual displays.

The configuration of the user interface is intended to address three questions:

1. Is the operating concept optimal from the point of view of road safety? Or in other words, can the driver find the information and operating element he needs immediately even in stressful situations?
2. Does the driver receive the information and control capabilities that support the most efficient and economical driving possible?
3. Is using the vehicle an easy experience that the driver enjoys or even looks forward to?

In order to lend some structure to the many displays and large amounts of information offered to the driver, many vehicles have two display areas. The primary display area is the standard main instrument panel, which is dedicated mainly to displaying information that is useful for the control of the vehicle during driving activities. This is located directly in front of the driver. There is often a secondary display area next to the primary display area, which places greater emphasis on information related to driver comfort and infotainment. This is where the operating information for the air-conditioning system, the audio equipment and navigation system usually appear. Figure 1.8 shows a modern HMI in a heavy truck with two fully digital displays. The content shown on the displays can be adjusted according to the specific usage.

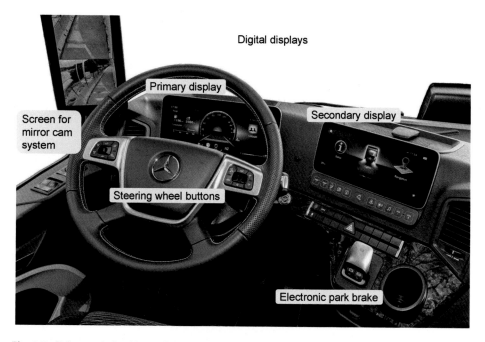

Fig. 1.8   Driver cockpit with two fully digital displays. (Photo: Daimler)

1.7.1 Main Instrument Panel

After direct sensory perception of the vehicle and its behavior (i.e. vehicle motion, ramps up speed, squeaks, rumbles, or stutters), the main instrument panel is the driver's primary source of information to support driving activities and the condition of the vehicle.

The vehicle speed display is usually located in the middle of the main instrument panel. Other visual measurement values that are often prominently shown are the engine speed, fuel tank level, the AdBlue or DEF—level and the pressure in the brake circuits. In some cases, there are laws in certain countries regarding what must be displayed. The main instrument panel also has an odometer.

The display can be presented in the familiar round gauges with a needle, with a linear bar indicator or in a number field.

Figure 1.9 shows a main instrument panel in a truck such as is commonly used in trucks.

Most instrument panels include a display above the basic equipment in which a menu can be opened using steering wheel buttons or other operating switches. These menus may include, for example, a trip odometer, a trip computer, maintenance information, wake-up functions and navigation displays. In some vehicles, the operating functions for the radio, auxiliary heater, telephone or level control are also included in the menu [16]. Maintenance information is often provided as well. Some digital menus are quite extensive, and in these cases marketing might refer to them as the onboard computer.

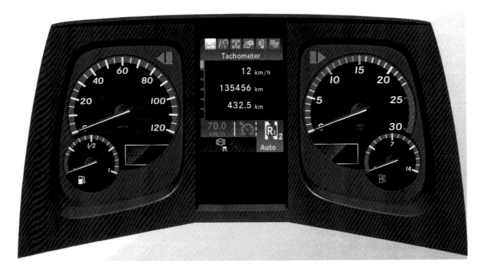

Fig. 1.9 Main instrument panel in a long-distance haul truck. The two large, round gauges show vehicle speed (left) and engine speed (right). The smaller round gauge on the *left* shows the fill level in the fuel tank, while the fill level of AdBlue or DEF is displayed in the smaller round gauge on the *right*. In the *center* is a LCD display in which many digital menus can be called up. (Illustration: Daimler)

Apart from the displays for continuous vehicle monitoring, the main instrument panel also has a number of indicator lamps that are displayed when certain functions are active. For example, there are indicator lamps for the turn signals, low beams, high beams or the retarder system. The indicator lamps light up when there is a need to communicate information to the driver; otherwise they remain dark and are not visible on the instrument panel. The instrument panel also includes functional status displays that alert the driver of malfunctions. These are usually assigned various levels of urgency. For example, high priority warnings are backlit in red, while lower priority warnings are yellow. Instruments with digital displays (as in Fig. 1.8 and Fig. 1.9) provide instructive texts regarding the malfunctions.

As the performance capabilities of displays have steadily increased, it is now possible to replace the standard instruments entirely with digital displays. These may display various types of information depending on the situation the vehicle happens to be in at the time. For example, a different display might be presented for highway driving than for maneuvering within city limits, or when the vehicle is stationary with the engine switched off, when infotainment and hours of service information may only be needed. Figure 1.8 shows fully digital displays.

Pneumatics System

Medium-duty and heavy-duty commercial vehicles (with a total vehicle weight of 7.5 t or more) are equipped with pneumatic brake systems.

Brakes actuated by compressed air have substantial advantages when the vehicle's brake system and the trailer's brake system are coupled together. A separable coupling with another medium such as hydraulic oil is always associated with the risk that hydraulic oil may leak during coupling and must be replenished, that it will cause soiling and that air may get into the system and impair the braking effect.[1]

If the medium of compressed air is already available because of the braking system, it makes sense to use it for other functions as well. The following are examples of other systems on the vehicle which are operated pneumatically:

- The actuation force for the clutch operating mechanism and the force for shifting operations in automated transmissions can be applied pneumatically.
- Many vehicles have air suspension.
- The comfortable seat in the cab is sprung pneumatically and can be adjusted with compressed air. It is called a suspension seat.
- Even the steering wheel adjustment is often powered pneumatically.
- The cab can be air-sprung.
- The parking brake is released with compressed air.
- The AdBlue for treating the exhaust gas can be injected with compressed air.
- Some of the functions of the truck body can also be actuated with compressed air.

---

[1] Historically, pneumatic brakes were first developed as a technology for railways. In railway applications, the coupling problem is many times more significant, because many carriages are coupled one behind the other.

© Springer-Verlag GmbH Germany, part of Springer Nature 2023
M. Hilgers, *Electrical Systems and Mechatronics*, Commercial Vehicle Technology,
https://doi.org/10.1007/978-3-662-66718-7_2

- Many trucks are equipped with a pneumatic horn in additional to the electric horn. The volume and tone of a pneumatic horn fit better to a heavy truck than an electric one.
- Vehicles are equipped with a connector for filling the tires with air.
- In some vehicles, the driver has a compressed air outlet that can be used for cleaning the vehicle with a compressed air gun.
- On buses, the doors are operated pneumatically.

The following components generate the compressed air for these systems:

Compressed air generation

The air compressor takes ambient air in, compresses it and discharges it into the compressed air system. Conventional air compressors are attached rigidly to the engine and deliver air while the engine is running. A pressure regulator therefore monitors the pressure in the compressed air system. When the supply pressure has been reached, the pressure regulator opens a port to the outside so that the displaced air is blown out with practically no resistance. The power consumption of the air compressor falls significantly, but the idle power of the compressor will still cause some fuel consumption.

The function of the pressure regulator is actuated mechanically. A spring force acts against the system pressure. When this reaches the shutoff pressure, the blow-off port opens. The pressure regulator can be a separate component, or it can be integrated as a function in the air dryer.

There are also designs with compressors that can be switched off. A coupling between the air compressor crankshaft and the diesel engine drive unit can be used to shut down the compressor completely. This reduces the vehicle's power draw and fuel consumption. Switching off the air compressor can also be expected to prolong the service life of the compressor. However, this advantage must be weighed against the additional component, namely the coupling. This additional component adds costs and extra weight and might fail.

The compressed air has to be dried to prevent moisture from accumulating, forming ice at low temperatures and clogging the compressed air lines. The air dryer purifies and dries the compressed air that is generated by the air compressor. Old compressed air concepts without an air dryer or with limited air drying capabilities include a device which adds antifreeze protection to the compressed air.

Distribution and storage of compressed air

The compressed air is distributed to various pneumatic circuits. Circuit 1 and circuit 2 serve the brake system (Chap. 3). Depending on the vehicle equipment, other circuits supply compressed air to the parking brake, the trailer, the air suspension, and for actuating the clutch and gearshift or similar. If the compressed air system is empty – after the vehicle has been out of use for a long period, for example – the two circuits for the service brake are always filled first for safety reasons. If there is a malfunction in either of the circuits, it must be ensured that the other circuits do not lose compressed air and remain operable. In traditional systems, this is done by the four-circuit protection valve. In modern vehicles an air processing unit (APU) combines several functions in one component. It dries the compressed air, regulates and distributes it to the individual circuits, and protects the entire compressed air system if a circuit leaks.

Compressed air reservoirs in each circuit create a reserve volume, so that even functions that require a large volume of air (e.g. lift vehicle) can be performed quickly. The compressed air reservoirs form a buffer volume that stabilizes the air pressure in the circuit when air is consumed. The size of the compressed air buffer in each circuit is set according to the vehicle's equipment configuration. For the brake system, there are legal regulations that specify the number of braking actions the buffer volume must be able to support if the supply of additional compressed air fails. Here the buffer volumes serves safety considerations as well.

The air pressures in the circuits are sensed and displayed to the driver on the instrument panel. A lamp warns the driver if the pressure of the compressed air is insufficient (when driving off, for example).

The lines for the entire compressed air system are routed in such a way that no water pockets can form, or dirt can collect in them. The compressed air reservoirs should ideally be located at the lowest point in the compressed air circuit, so that any moisture and dirt in the circuit collects there.

If the air dryer function is inadequate or even defective, condensation can collect in the compressed air reservoir. This is why the compressed air reservoirs require drain valves. The driver must actuate them before starting the vehicle in order to gauge if there is any condensation in the reservoirs.

A condensation sensor may be installed to monitor the formation of condensation in the compressed air system and reduce the manual effort on the driver. The sensor detects the condensation, and a display informs the driver that condensation is forming in the compressed air reservoir. If a significant quantity of condensation forms, the compressed air system must be checked for leaks or malfunction of components (airdryer).

Brake System

3

The braking function fulfills three tasks: It serves to reduce the speed of the vehicle (i.e. decelerates it more rapidly than the effect of ubiquitous friction alone), it helps to control the vehicle's speed when traveling downhill, and it keeps the vehicle at rest when it is stopped or parked.

A distinction is made between the service brake, the permanent brake (retarder) and the parking brake. The function of the service brake is to enable the vehicle's speed to be reduced gradually while driving. The service brake is usually actuated by the brake pedal. The permanent brake is an additional brake system that can brake for a prolonged period without noticeable loss of effect; the permanent brake is usually a retarder or a particularly powerful engine brake. The driver usually operates the permanent brake with a handlever. In some older vehicles, the permanent brake is operated by foot. If the driver wishes to slow or stop the vehicle while in motion, the service brake and the optional permanent brake system can be utilized. Modern brake systems do brake integration. The driver's intention to decelerate is evaluated electronically and the control algorithm in the vehicle decides how the intended deceleration is to be carried out with a combination of wheel brake and permanent brake actuation. The permanent brake system supports the service brake function in this manner.

The parking brake is responsible for holding the vehicle when it is stationary on level grounds as well as on inclines. The parking brake is usually operated by hand. Requirements on brake systems differ for various vehicle categories. Table 3.1 shows the vehicle categories used in European regulations.

The secondary braking system ensures that the vehicle can be slowed down (with a lower deceleration) if the service brake fails. The brake system is configured with two circuits. If one brake circuit fails, the vehicle can still be slowed down by the second circuit. The braking system must still be able to stop a vehicle when there is a single point of failure in the service brake system, including loss of air. As a safety–critical function, the brake must be compliant with many regulations, for example, it must still be possible

© Springer-Verlag GmbH Germany, part of Springer Nature 2023
M. Hilgers, *Electrical Systems and Mechatronics*, Commercial Vehicle Technology,
https://doi.org/10.1007/978-3-662-66718-7_3

Table 3.1 Commercial vehicle categories according to Annex II of Framework Directive 2007/46/EC (replaces 70/156/EEC)

	M_1	M_2	M_3	
Category M: Vehicles designed and constructed for the carriage of passengers with at least four wheels	Driver plus up to 8 seats	Driver and more than 8 seats, up to 5 t	Driver and more than 8 seats weighing more than 5 t	

	N_1	N_2	N_3	
Category N: Vehicles designed and constructed for the carriage of goods with at least four wheels	Having a maximum mass not exceeding 3.5 t	3.5 to 12 t	More than 12 t	

	O_1	O_2	O_3	O_4
Category O: Trailers	With a maximum mass not exceeding 0.75 t	0.75 to 3.5 t	3.5 to 10 t	More than 10 t

to apply the brakes a defined number of times even after the energy supply fails, or the supply of compressed air is not sufficient. The regulatory requirements for brake systems in the United States are set forth in FMVSS 121 [25]. In Europe, the regulatory requirements for the different vehicle classes are given in ECE-R 13 [18].

A distinction is made between power-assisted brake systems and external power brake systems. Power-assisted brake systems assist the driver's physical strength, either pneumatically (i.e. vacuum brake booster) or hydraulically. If the pneumatic or hydraulic assistance fails, the driver can still rely on muscle power to apply the brakes. With external power systems, the driver's strength adds nothing to the braking force. Actuation by the driver simply controls the brake. Larger commercial vehicles typically use external power systems while cars and vans have power-assisted brake systems.

The purpose of braking is to deplete the vehicle's kinetic energy W_{kin}. This is done by converting the kinetic energy into heat. The kinetic energy is calculated using m for mass and v for speed (velocity) in the following equation[1]:

$$W_{kin} = \frac{1}{2}mv^2 \tag{3.1}$$

For a mass m = 40 t or 40,000 kg and a speed of 90 kph or 25 m/s, this corresponds to 12.5 MJ of energy. If the vehicle is decelerated at a rate of 6 m/s² with emergency braking, the calculated braking time is somewhat above 4 s. The average heat output during

---

[1] Energy and work are often denoted with W for work.

Table 3.2 Coefficients of
friction between tires and road
for selected road types

Road surface	Dry	Wet
Ideal concrete	≈1	
Concrete	0.6–0.9	0.4–0.7
Asphalt	0.6–0.8	0.3–0.6
Dirt roads	0.4–0.5	0.3
Blue basalt[a]		0.15
Snow	0.2	0.1
Stainless steel (wet)		0.1
Black ice	0.1	0.01–0.1

[a]Blue basalt is a commonly used road material for test surfaces because it can be used in the wet condition as a substitute for a road covered with compacted snow. The blue basalt road surface is not used in dry conditions

that time is about 3 MW.[2] For the sake of comparison, the central heating system in a modern single-family home may perhaps generate up to 10 kW. This means that when the brakes on a 40-t truck are applied fully, for a short time the wheel brakes generate as much heat energy as 300 detached houses! At least for a short period, the brake system can convert about 6–10 times as much output as the engine, the maximum output from which is in the order of 300 to 500 kW. But the brake can only process this amount of energy briefly. The brake gets very hot and, in the process, loses its effectiveness. This phenomenon is called brake fade.

The maximum possible deceleration a_{max} of a vehicle depends on the coefficient of friction between the tires and the road μ_k:

$$a_{max} = \mu_k \cdot g \tag{3.2}$$

g^3 is the standard acceleration of gravity, about $g = 9.81$ m/s².

The coefficient of friction is a dimensionless variable that describes the strength of the friction between the vehicle's tires and the road. Table 3.2 shows a few examples of the friction coefficient μ_k. With a friction coefficient of 1, a vehicle can be decelerated by braking at a rate of about 10 m/s². For commercial vehicle-tires on a high-friction surface, the friction coefficient is in the order of $\mu_k = 0.8$ so we can expect a deceleration of around 8 m/s² under ideal conditions.

The maximum braking deceleration achievable is obtained using the maximum friction of the tires on the road surface. The braking deceleration depends on weather

---

[2]The equations you need here are as follows, where t stands for time, v for speed and a for acceleration $t = v/a$ and $P = W/t$ with P as output.

[3]The normal g force varies slightly depending on the geographical location.

conditions and the condition of the road, and also on the condition of the tires. If the braking system is operated too aggressively, the wheels may lock. The friction pairing between tires and road is changing from static friction into sliding friction. If the front axle locks, the vehicle slides and no longer responds to steering commands. If the rear axle locks, the vehicle becomes unstable and skids.

When the performance of a braking system is discussed, the term braking ratio is often used. The braking ratio z is the ratio between the braking force of the vehicle F_{Brake} and its weight force F_G Eq. 3.3:

$$z = \frac{F_{brake}}{F_G} \tag{3.3}$$

The braking ratio is typically expressed as a percentage.

The braking force takes effect on the underside of the tires, way below the vehicle's center of gravity. As a result, when the brakes are applied, torque is generated about an axis transverse to the direction of travel (usually described as the y-axis), additional load is transferred to the front axle and the rear axle(s) are relieved, so the vehicle tilts forward. This is called dynamic axle load transfer.

3.1   Wheel Brake or Foundation Brakes

The purpose of the foundation brakes is to convert the vehicle's kinetic energy into heat. In modern commercial vehicles, two wheel brake technologies may be implemented: drum brakes and disc brakes. Both operate according to the same basic principle: One component, called the brake rotor, the drum in the drum brake and the disc in the disc brake, rotates with the wheel. When the brakes are applied, the unit that is fixed on the axle forces the brake linings against the rotating component. The generated friction causes the vehicle to slow down and converts kinetic energy into heat. This causes abrasion of the brake lining, and to a lesser degree, of the drum or disc as well. These parts are replaced if worn out.

Figure 3.1 shows how the categories of wheel brake systems are related.

Important criteria for the design of the wheel brake are wear, mechanical strength, thermal resilience, heat absorption capability or heat conductivity, and heat dissipation, as well as brake cooling capability. The drum brake and the disc brake each have specific advantages, so it is quite understandable that both wheel brake designs are still in use. However, as a long-term trend, disc brakes are increasingly gaining favor over drum brakes. Historically, one advantage of the drum brake has been the internal gain, or self-servo action. This means that during the braking operation the contact pressure of the brake linings against the drum is reinforced. This inner reinforcement is described with the brake factor C*. Depending on design type, the brake factor exerted by a drum brake can be as much as five times stronger than with a disc brake. Modern braking systems have so much wheel brake actuating force at their disposal that this self-servo braking

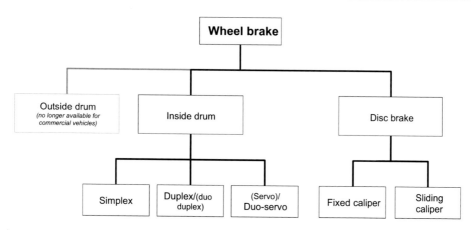

Fig. 3.1 In modern commercial vehicles, drum brakes and disc brakes are used

effect is irrelevant; in fact, it is even disadvantageous. A brake with inherent reinforcement cannot be applied in such precise increments. It responds more aggressively to small changes that occur over the service life of the brake. The disc brake is less sensitive to fluctuations in the friction coefficient and so a disc brake system is less prone to pulling the vehicle to one side while braking ("skewing"). The disc brake can also be cooled more efficiently, so more braking power can be generated in the same installation space. Furthermore, there is no fading with a disc brake. Fading refers to the phenomenon in which the braking performance diminishes as the brake becomes hotter. Fading occurs in drum brakes when heat causes the drum to expand outwards. This in turn reduces the friction between the brake lining and drum. Because of its geometrical configuration, the disc in a disc brake does not exhibit this effect when heated.

One advantage of the drum brake, which is still significant, is its closed construction. This makes it easier to protect the brake effectively from soiling.

3.1.1 Drum Brake

Drum brakes with brake shoes that are pressed against the drum from inside have been in use for more than a hundred years.

Simplex brake
The simplest basic form of the drum brake is the simplex brake. The two brake shoes are mounted rotatable at one end; and at the other end during application of the brakes they are moved apart and against the drum—see Fig. 3.2.

The friction between the brake lining and the drum generates a tangential force that causes the brake shoe on one side to be pressed more firmly against the drum. This is the leading brake shoe (or primary shoe). The actuating force on the leading brake shoe

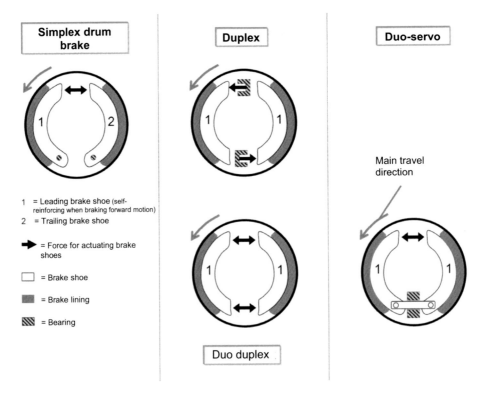

Fig. 3.2 Schematic representation of the different drum brake design types. Explanatory details are given in the text

is amplified. On the other brake shoe, the tangential friction counteracts the actuating force. This is the trailing or secondary shoe. Consequently, the braking effect is initially stronger with the leading shoe.

If the brake actuation unit is floating (i.e. the brake actuation unit can move horizontally and the two brake shoes brace each other mutually) the leading brake shoe permanently contributes more to the braking effect, and the brake lining therefore wears more quickly than the lining on the trailing shoe. This effect can be addressed by using linings of different thicknesses on the leading and trailing brake shoes. If the brake actuating unit is on a fixed mounting (as with the S-cam shown in Fig. 3.3), the rates of wear on the brake linings of the leading and trailing brake shoes automatically converge. Initially, wear is slightly more pronounced on the leading brake shoe, during subsequent braking actions, this brake lining is pressed less strongly against the drum than the thicker lining of the trailing shoe. As a result, the two linings ultimately wear at approximately the same rate.

When the vehicle is reversing, the roles of the two brake shoes are also reversed. The leading and trailing shoes still exist, but they are performed by the other shoe. The braking effect with a simplex brake is the same for forward and reverse travel.

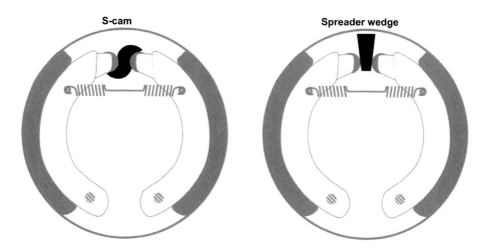

Fig. 3.3 Principle of the spreader wedge and the S-cam for spreading brake shoes in a drum brake. The S-cam rotates and forces the brake shoe outwards. The spreader wedge moves linearly inwards, forcing the brake shoe outwards

There are various ways to spread the brake shoes. With hydraulic brakes in cars or vans, there is a hydraulic cylinder. In the pneumatic brake of a truck, the pneumatic pressure is converted into the movement of a piston rod in the brake cylinder (see that section for more details). The piston rod moves a spreading wedge or the S-cam, either of which then displaces the brake shoes in the drum brake. The principles of the spreading wedge and the S-cam are illustrated in Fig. 3.3.

Drum brakes on trucks are usually designed as simplex brakes.

Duplex brake and duo duplex brake
The duplex brake provides independent operation for each brake shoe (see Fig. 3.2). Two actuating mechanisms are needed in the brake drum, so this makes it much more complicated. The two independent actuating mechanisms in the duplex brake lend both brake shoes an inner reinforcement and a higher total brake factor than for the simplex brake. The braking effect is more powerful for the same control force. When the vehicle is reversed, both brake shoes function act as trailing shoes and the braking effect in reverse is considerably weaker than for forward travel, and is also weaker than with the simplex drum brake.

This disadvantage of low brake power in reverse is corrected with the duo duplex brake. Here, the two actuating mechanisms press both brake shoes outwards. The brake is perfectly symmetrical, so the same braking power is applied in both forwards and reverse. The wear on the brake linings of both brake shoes is also symmetrical.

Duo-servo drum brake
The duo-servo drum brake is also symmetrical. The brake shoes are mounted on a floating caliper. The first leading brake shoe is braced on the second brake shoe by a bolt,

and this turns the second brake shoe into a leading shoe as well. The inner reinforcement effect in this arrangement is very effective, and relatively little operating force needs to be applied. On the other side, however, the brake cannot be applied with the same degree of precision. The duo-servo brake behaves in the same way for forward and reverse travel.

3.1.2 Disc Brake

In disc brakes, the brake disc is attached to the wheel hub and rotates with it. The brake caliper does not rotate; and when the brakes are applied it forces the brake linings against the disc. There are two kinds of disc brake: the fixed caliper brake and the sliding caliper brake. In the fixed caliper brake, the caliper is mounted rigidly and does not move when the brakes are applied. In this case, actuators are needed on each side of the brake disc to press the brake linings against it.

The sliding caliper brake relies on the Newtonian principle of action and reaction. The brake caliper is mounted in such a way that it can be shifted perpendicularly to the brake disc. An actuator on one side of the brake disc, presses against the disc. The counterforce causes the caliper and the brake lining on the other side to shift, moving it towards the brake disc. The two brake linings act symmetrically on the brake disc. Figure 3.4 shows a diagram of a sliding caliper disc brake. When filled with compressed air, the

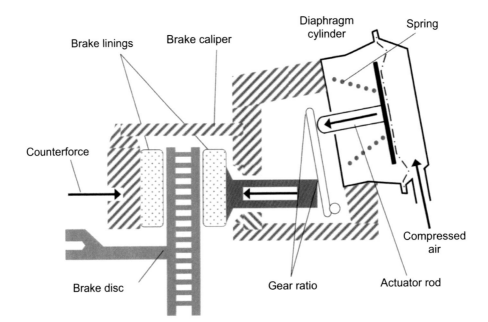

Fig. 3.4 Schematic representation of a sliding caliper disc brake

diaphragm cylinder extends the actuator rod. This rod actuates a lever (that usually adds a lever ratio/amplification) in order to force the brake lining against the disc. The internal mechanism of the disc brake with the amplification ratio is highly simplified here for the sake of better comprehensibility. Under the effect of the counterforce, the entire brake caliper is shifted together with the diaphragm cylinder and also presses the other side of the caliper against the disc with the same force. The force with which the brake caliper is applied depends on the pressure built up in the diaphragm cylinder, the effective area of the diaphragm cylinder and the amplification ratio of the mechanism in the brake.

3.2   Brake Cylinder

The brake cylinder converts pneumatic pressure into mechanical movement on the foundation brake. The actuator rod of the brake cylinder moves. In brake systems with disc brakes, the brake cylinder rod can act directly on the foundation brake (see Fig. 3.4). In brake systems with drum brakes, a linkage is responsible for ensuring that the movement of the brake cylinder actuator rod is translated into rotary motion of the S-cam or linear motion of the wedge.

The force with which the rod is moved depends on the pneumatic pressure supplied and the surface area of the brake cylinder piston: Force equals pressure times area, $F = p \cdot A$. There are a number of different brake cylinders with various piston sizes.

Axles, which are braked only by the service brake, are equipped with diaphragm cylinders. Axles with a parking brake as well as a service brake are equipped with a combination cylinder. The combination cylinder combines the spring loading of the parking brake (or handbrake) and the functionality of the diaphragm brake cylinder. The combination cylinder will be explained below. If you understand the combination cylinder, you understand the simple diaphragm cylinder consequentially.

Figure 3.5 shows how a combination cylinder works. The right side of the combination cylinder is what is known as the diaphragm cylinder. On the left is the spring-loaded brake actuator for the parking brake. In the park position shown in Fig. 3.5a, the spring-loaded brake actuator and the diaphragm cylinder are not supplied with compressed air. The spring in the spring-loaded brake actuator (1) presses the piston (2) forwards and the push rod (3) with it. The spring-loaded brake actuator overcompresses the spring in the diaphragm cylinder (5) so that the actuator rod (6) is pushed outwards and the brake is applied.

To release the spring-loaded brake actuator i.e. the parking brake, the chamber (x) is filled with compressed air. See representation in Fig. 3.5b. Piston 2 works against the spring of spring-loaded brake actuators (1) to force the piston rod (3) back. The spring of the diaphragm cylinder (5) pushes the actuator rod (6) back and the brake is released. While the vehicle is being driven, the chamber (x) is permanently pressurized with compressed air.

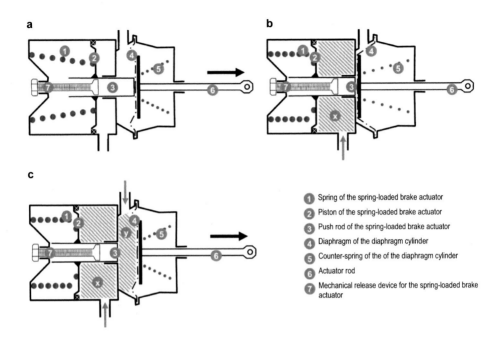

Fig. 3.5 Schematic representation of a combination brake cylinder: The *right* side of the combination brake cylinder is the diaphragm cylinder, the *left* side is the spring-loaded brake actuator unit (park brake functionality). Axles that are not controlled by the parking brake are equipped only with diaphragm cylinders. The left side of the combination cylinder is not needed (**a** Park position, **b** Released position, unbraked, **c** Braking with the service brake). The functionality is explained in the text

When the service brake is activated, compressed air flows into chamber (y)—Fig. 3.5c. This causes the diaphragm (4) of the diaphragm cylinder to move against the spring force of spring (5), pushing the actuating rod (6) out. The vehicle is then braked. After braking with the service brake has ended, the compressed air is released from the chamber (y) and spring (5) retracts the actuator rod (6) again.

The bolt (7) can be used to release the parking brake (compress the spring of the spring-loaded brake actuator) when there is no compressed air in the vehicle. This is necessary, for example, if a broken-down vehicle must be towed. The bolt (7) is unscrewed. It brings the push rod (3) with it, so the piston (2) is moved against the spring (1), and the actuator rod (6) is retracted by the spring force of the spring (5).

If the parking brake and the service brake are applied at the same time, the braking force in the combination cylinder is increased further. In certain circumstances, this can lead to mechanical overload of components of the braking system. Therefore, overload protection is needed. In an electronic brake system, the maximum braking pressure supplied to the service brake is limited electronically when the parking brake is engaged. In a purely pneumatic brake system, the anti-compounding function is assured with a valve.

3.3   Parking Brake or Handbrake

The task of holding the vehicle still when it is stationary falls on the parking brake, also commonly called the handbrake. This too is subject to specifications according to FMVSS 121 and ECE-R 13 respectivly. For example, the parking brake system must be able to hold the loaded vehicle on an specified grade (18% with ECE-R13 and 20% with FMVSS 121). ECE-R 13 requires on top that the loaded vehicle with loaded trailer can be held on a 12% grade. The parking brake system must enable the driver to check whether the parking brake of the tractor vehicle is sufficient on its own to hold the vehicle combination on the incline even if the trailer brake is inoperative—this is done by the so-called check position on the hand brake actuation lever.

When the spring-loaded cylinder has been depressurized (no air in it), the brake is in the park position. In the drive position, the spring-loaded brake actuator is pressurized (filled with air). The parking brake is actuated by the parking brake valve. Normally, the driver operates the parking brake valve directly with the handbrake lever in the cab, and this then fills the spring-loaded actuator with compressed air or expels the air as explained in the previous Sect. 3.2. There are also some vehicles with electric handbrake operation—see Chap. 1 Fig. 1.8. The driver presses an electrical switch to trigger a signal which then initiates actuation of an electromagnetic handbrake valve outside of the cab. The advantage of the electric parking brake valve is that pneumatic lines no longer have to be routed from the cab to the valves for the parking brake function. Moreover, an electronically actuated handbrake opens up the possibility of designing other functionalities that the solely pneumatic parking brake could only fulfill with a great deal of difficulty, if at all. For example, with an electric parking brake, it is possible to release the parking brake automatically when driving off.

The compressed air used to release the parking brake is drawn from a compressed air circuit which is subordinate to the service brake circuits 1 and 2. Since circuits 1 and 2 must always be filled first for safety reasons, the parking brake cannot be released until the service brake circuits have been filled with air and the vehicle can therefore be held immobile by the driver using the service brake.

An indicator lamp on the instrument panel informs the driver when the parking brake is engaged.

3.4   Pneumatic Brake System

We start this section with a description of the basic principle of the purely pneumatic service brake of a two-axle vehicle. The real brake system is typically much more complex than this. The reader will realize the truth of this later when additional functions are described.

Figure 3.6 outlines the basic layout of the pneumatic brake system for a truck. A compressed air circuit supplies the actuating force for the rear axle (circuit 1) and a second compressed air circuit serves the front axle (circuit 2). Two valves, each operated by the brake pedal, allow compressed air to flow into the brake cylinders. The brake cylinders convert the pressure of the compressed air into a mechanical movement, which actuates the wheel brake. In this simple brake, the characteristic curve that represents the ratio of brake pedal travel to braking force is set by the mechanical valve actuation via the brake pedal. This brake system operates solely by pneumatic mechanical action.

If an anti-lock braking system (ABS) is integrated in the braking system, additional pneumatic and electronic components are required.

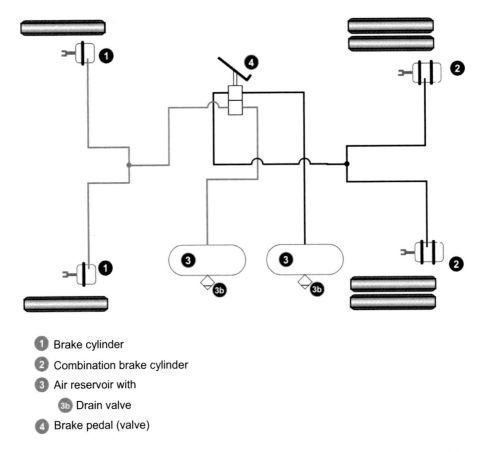

① Brake cylinder

② Combination brake cylinder

③ Air reservoir with

    3b Drain valve

④ Brake pedal (valve)

Fig. 3.6 Pneumatic diagram of the basic layout of a brake system for a truck with no additional functions such as ABS, etc. For the sake of simplicity, the compressed air supply is not shown

3.5   Anti-Lock Braking System—ABS

It is possible that the wheels may lock when braking. When the wheels lock, the vehicle loses its tracking stability and is in danger of skidding. It is also impossible to steer a vehicle with locked wheels because the steered wheel that is locked can no longer deliver lateral stability. Furthermore, it is usually impossible to guarantee the shortest braking distance when the wheels are locked. The anti-lock braking system (ABS) controls the brake force electronically such that maximum deceleration can be carried out without having a wheel stop turning, in other words locking. To do this, the angular velocities of the wheels are monitored by rpm sensors (rotational speed sensors). The wheels are kept continually just below the locking threshold by the cyclic dissipation and buildup of brake force. The vehicle remains steerable, and the best possible braking distance is achieved.

Figure 3.7 shows the layout of an ABS system. The sensor and the pole wheel measure the angular velocities of the wheels and transmit this information to the ABS ECU.

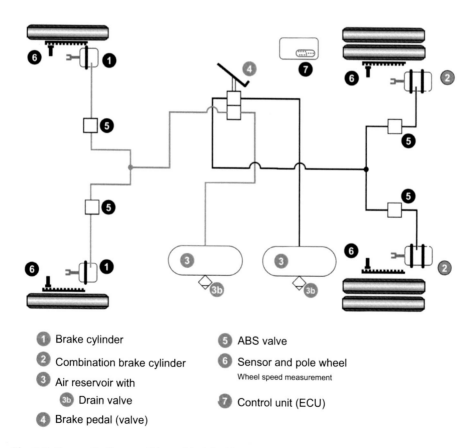

1 Brake cylinder

2 Combination brake cylinder

3 Air reservoir with
  3b Drain valve

4 Brake pedal (valve)

5 ABS valve

6 Sensor and pole wheel
  Wheel speed measurement

7 Control unit (ECU)

Fig. 3.7 Pneumatic diagram of the anti-lock braking system (ABS)

The ECU determines how strongly each wheel must be braked. The ABS valves are controlled individually for each wheel. The brake pressure can be increased, reduced or held constant at the existing brake pressure for each wheel individually.

Figure 3.7 shows the ABS system with two ABS valves for the rear axle. There are also configurations in which both sides of the rear axle are controlled via one valve. The configuration with two rear-axle ABS valves is needed for a brake with a traction control system (ASR or TCS).

3.6   Traction Control, Traction Control System (TCS)

On a slippery surface, the risk exists that the driven wheels of the vehicle may slip. The tendency of the drive axle to spin uncontrollably increases significantly on slippery surfaces, particularly in unladen commercial vehicles. Firstly, this makes travel difficult—in extreme cases on a slippery road it can make launching completely impossible—and secondly, spinning wheels lose their ability to deliver lateral stability. So a vehicle with too much slip on the drive axle is less stable. Furthermore, the tires wear faster.

The purpose of traction control or anti-slip regulation (ASR) is to prevent drive wheels from slipping. In order to do this, the rpm information from the wheel sensors is evaluated in the ECU. At first, the driving wheel on one side usually starts slipping. But because of the axle differential in the axle, this slipping wheel is then responsible for the loss of propulsive power that affects the entire vehicle. ASR slows down the spinning wheel by applying the brakes, so the wheel with better traction (grip) can accelerate and move forward. To do this, the ASR control directs brake pressure to the drive axle brake circuit by means of the ASR valve. The two-way valve in the ASR system delivers either the brake pressure activated by the driver or the brake pressure delivered by the ASR valve—whichever is greater—to the brake circuit of the traction axle. A single wheel is then braked individually through the ABS valves on the drive axle. Figure 3.8 shows the brake system with ASR components.

Besides the braking action of the ASR system, acceleration skid control can also act on the engine control. If the driver accelerates too sharply and both wheels slip, the engine torque is reduced to restore the stability of the vehicle and bring the tire slip back to a value which is suitable for good traction. At high speeds, the ASR system only intervenes via the engine control and does not use the brake system to avoid excessive load on the brakes.

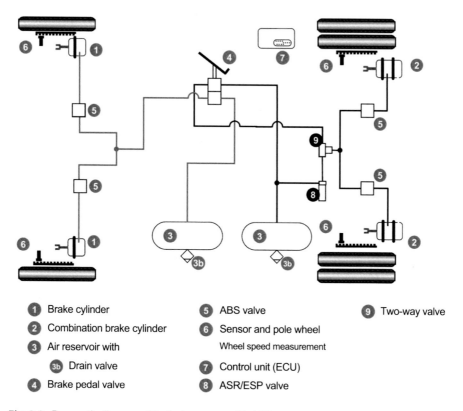

① Brake cylinder   ⑤ ABS valve   ⑨ Two-way valve

② Combination brake cylinder   ⑥ Sensor and pole wheel

③ Air reservoir with   Wheel speed measurement

     ③b Drain valve   ⑦ Control unit (ECU)

④ Brake pedal valve   ⑧ ASR/ESP valve

Fig. 3.8 Pneumatic diagram of the brake system with ASR

3.7 Electronic Stability Program—ESP

The electronic stability program (ESP)—sometimes the abbreviation ESC is used—serves to improve track stability and secondly, the roll stability of the vehicle.

Improvement of stability means that within the limits of the physically possible, ESP prevents the vehicle from skidding, or the truck trailer combination from jackknifing.

To do this, the ECU of the ESP evaluates the rpm sensors, which measure the rotational speed of each wheel individually. Additionally, the steering angle of the vehicle is detected and this is used to capture the travel direction desired by the driver. A sensor for lateral acceleration captures the vehicle's drift velocity, and the yaw rate sensor measures whether and to what extent the vehicle is rotating about the vertical axis (i.e. z-axis).

By evaluating the data from the various sensors, an algorithm determines whether the vehicle is obeying the driver's steering commands or whether it is about to swerve. With the aid of precisely coordinated brake applications, the ESP stabilizes the vehicle and steers the vehicle in the direction that matches the driver's steering command. The basic principle of an ESP system is to slow any wheel on the vehicle individually. This enables a turning moment about the vertical axis to be generated precisely, and the vehicle is steered. In the case of oversteer, ESP brakes the front wheel on the outside of the curve and straightens the vehicle. In the case of understeer, the rear wheel on the inside of the curve is braked, which creates a more pronounced turning movement into the curve. The ESP can also slow the trailer down in specific manner, which has the effect of stretching the tractor trailer combination and stabilizing it. In Fig. 3.9 the wheels which are braked in the control intervention are shaded gray.

ESP increases the truck's roll stability because the ESP electronics permanently monitor its lateral acceleration. If critical lateral acceleration values are reached, the vehicle is braked.

ABS and ASR form an integral part of the ESP system.

To enable the ESP system to individually brake not only the wheels on the drive axle but also the wheels on the steering axle, the same two valves are needed on the steering axle as are used in the ASR system for the driven rear axle. One valve serves to direct brake pressure to the brake circuit for the front axle, a two-way valve has the task of permitting either the brake pressure activated by the driver or the brake pressure delivered by the ESP valve—whichever is greater—to the brake circuit of the front axle. The individual braking of a single wheel then takes place via the ABS valves. A steering angle sensor and a sensor for measuring lateral acceleration and yaw rate are also needed. The components of an ESP brake system for a two-axle truck are shown in Fig. 3.10.

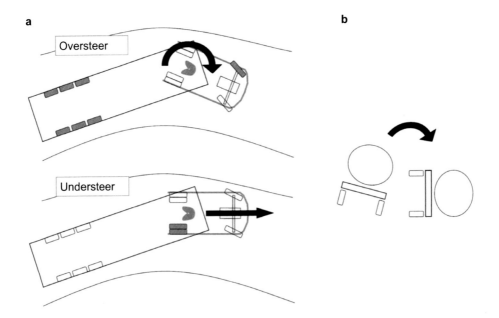

Fig. 3.9 Operating cases of the electronic stability program (ESP) system with the example of a tractor trailer combination. The wheels shaded gray are braked in a control intervention function. **a** Sway control, **b** Roll stabilization

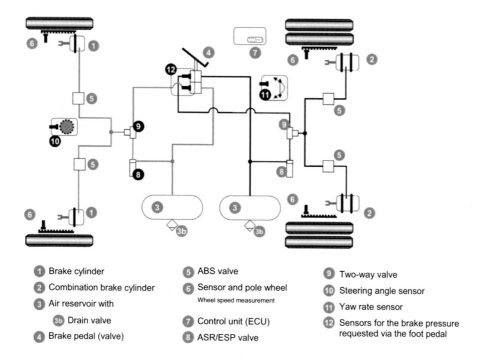

1 Brake cylinder		**5** ABS valve		**9** Two-way valve
2 Combination brake cylinder		**6** Sensor and pole wheel		**10** Steering angle sensor
3 Air reservoir with		Wheel speed measurement		**11** Yaw rate sensor
3b Drain valve		**7** Control unit (ECU)		**12** Sensors for the brake pressure
4 Brake pedal (valve)		**8** ASR/ESP valve		requested via the foot pedal

Fig. 3.10 Schematic representation of a brake system with ESP for a two-axle vehicle

3.8   Braking the Trailer

Trucks are typically designed to pull trailers. In order to supply the pneumatic brake information to the trailer, the tractor vehicle is equipped with the trailer control valve (Fig. 3.11). The trailer control valve delivers the brake pressure from both service brake circuits and the parking brake to the trailer. The pneumatic connection from the trailer control valve towards the trailer are the supply line (i.e. red coupling head) and the brake line (i.e. yellow coupling head). There are also coupling systems in which the supply line and the brake line are combined in one coupling, so that only one—albeit rather larger—coupling head—has to be manipulated when coupling or uncoupling the trailer.

A two-circuit brake system is required by law for the trailer as well. The distribution of the brake force to the vehicle axles and the division of force between tractor and trailer is specified in Annex 10 of ECE R-13 [18]. For the US, different regulations apply, hence trailer in the US and Europe differ.

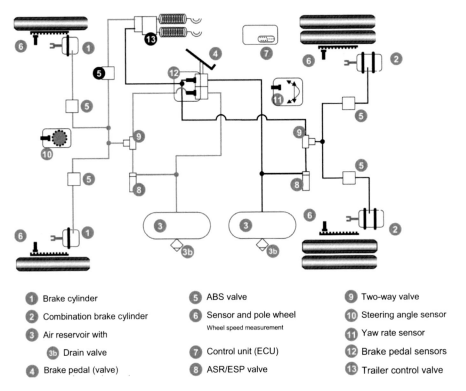

①	Brake cylinder	⑤	ABS valve	⑨	Two-way valve
②	Combination brake cylinder	⑥	Sensor and pole wheel	⑩	Steering angle sensor
③	Air reservoir with		Wheel speed measurement	⑪	Yaw rate sensor
3b	Drain valve	⑦	Control unit (ECU)	⑫	Brake pedal sensors
④	Brake pedal (valve)	⑧	ASR/ESP valve	⑬	Trailer control valve

Fig. 3.11 The trailer control valve forwards the braking command from the vehicle to the trailer or semitrailer

3.9   Other Brake Functionalities and Components

Other components that are essential for some commercial vehicle brake systems are listed below.

Automatic load-dependent brake force distribution: Vehicles that do not have an electronic control for the brake force distribution are equipped with the automatic load-dependent brake (ALB) valve for automatic load-dependent brake force distribution. This controls the brake pressure applied to the rear axle depending on the loading of the vehicle. If the vehicle is empty, the brake pressure on the rear axle must be lower to avoid overbraking the rear axle. As the load on the rear axle increases, the axle is able to take on a larger share of the braking action. On vehicles with steel suspension, the ALB valve can be controlled purely mechanically via a linkage. On vehicles with air suspension, the pressure from the air spring bellows is used to actuate a pneumatically controlled ALB valve. The ALB valve is no longer necessary in modern brake concepts. The distribution of brake pressure to the axles is calculated on the basis of a large volume of information in the ECU in order to achieve the desired incremental deceleration with the greatest possible stability (i.e. safety) and comfort.

The brake systems in some vehicles are equipped with the **auxiliary brake valve**. In normal operation, the auxiliary brake valve has no effect on the application of the service brake. Its only purpose is to ensure sufficient deceleration in the event that the front axle brake circuit fails, because there are some vehicles for which the braking effect of the rear axle circuit alone is not sufficient if the pneumatic circuit to the front axle fails with unfavorable load conditions. In these vehicles, the auxiliary brake valve directs some of the compressed air from a third circuit to one of the two front wheels, so that this wheel is partially braked and in an emergency adds to the overall braking effect on the vehicle.

The **quick-release valve** is used to vent the brake cylinders and pneumatic control lines more quickly so that the brake can be released sooner. **Relay valves** serve to fill a compressed air volume faster and then vent it (i.e. expel the air again faster). They fulfill a rapid release and a rapid filling function in one. **Pressure limiting valves** are used to limit the pressure when parts of the compressed air system are working under reduced pressure. Other components that may be found in brake systems include overflow valves, check valves, brake line filters and brake line fracture protectors.

3.10 Electropneumatic Brake

In principle, the pneumatic brake that has been described in the preceding sections is still the brake of the modern truck. Even so, the pneumatic operating principle is now supported by an electronic control. This is called the electropneumatic brake (EPB) or the electronic brake system (EBS).

With the EPB, signals are transmitted electronically in normal, fault-free braking. For example, the braking signals are transmitted electronically, and the valves are actuated electronically. In normal mode, the vehicle functions in brake by wire mode. The pneumatic system is used to supply the actuation force but not used to transmit the braking signals. The conventional, dual circuit pneumatic basic brake system is subordinate to the ECU. The purely pneumatic functional capability of the brake system is retained. If the electronic actuation of the brake system fails for any reason, braking is assured in the standard way by means of the purely pneumatic functionality.

The brake value sensor (i.e. brake pedal) is equipped not only with the pneumatic control valve as shown in Figs. 3.6, 3.7, 3.8, 3.10 and 3.11, it also includes a sensor for the pedal position. The signal from the brake pedal is sent to the brake ECU. This unit registers the driver's intention and initiates braking. The valves at the wheel are actuated electronically to trigger braking (or the end of braking). The EPB initiates the start of braking faster and enables more precisely controlled braking. With the EPB it is also possible to implement additional functions (see [20]).

With the EBS it is logical, but not absolutely necessary to integrate the permanent brake (i.e. brake integration). It is decided by the electronics whether the service brake or the permanent brake or both are to be engaged to fulfill the desire to slow the vehicle down.

The EBS or EPB processes other electronic variables such as wheel speeds. The variable slip of the axles is calculated from the differing wheel speeds. With this information it is possible to determine instantaneously which axle can contribute how much to the braking effect. In this way, the optimum brake pressure per axle is determined based on the load and the condition of the brake and tires. This is called **differential slip control** by the EPB [19] and [20]. The vehicle is braked effectively and with the maximum possible driving stability. When calculating the ideal brake pressure for each axle, the total mass of the vehicle, the mass distribution, the road gradient and the friction value of the wheels on the road are all taken into account.

The EPB can also perform **brake wear harmonization**: Within certain limits, the EPB is able to adjust the distribution of brake force between the axles in partial braking, thereby harmonizing the wear on the brake linings of the axles. It is obviously very helpful for the customer if the brake linings undergo wear at approximately the same rate.

The EPB has an electronic interface, via which certain advanced driver assistance systems (See Sect. 4.1) can request a brake manouever.

From Advanced Driver Assistance Systems to Automated Driving

4

Advanced driver assistance system (ADAS) is the term used to refer to those systems which help the driver either by taking over a task the driver would otherwise have to carry out manually, or by fulfilling a function which is beyond the capabilities of the driver. These systems are predominantly aiming to increase safety. One of the great advantages of assistance systems is that, unlike the driver, they never get fatigue and cannot be distracted. A frequent cause of accidents is human error, and human error is often due to fatigue or distraction (smartphone!).

Long term more and more advanced driver assistant systems with more and more advanced capability point into the direction of autonomous driving covered at the end of this chapter. Autonomous driving will increase road safety (otherwise it will not be accepted by society) and will change the cost structure of freight forwarding companies. A lot of very reliable and versatile functions will be required to approach autonomous driving. Albeit some of today's ADAS functions will be no longer required like for example drowsiness detection: The autonomous system does not get tired.

Usually six levels of automation for road going vehicles are distinguished from level 0—no automation, everything is done by the driver—to level 5—autonomous truck: no driver needed. This classification goes back to a SAE standard [26] and is widely accepted and taken over by authorities [27]. A similar classification is used in other industries like e.g. agricultural machines [28]. Figure 4.1 shows the 6 levels, explains some details and gives examples. With each level the system takes over more and more tasks from the driver.

To realize automated systems the vehicle must have an environment recognition (see later section) which uses sophisticated sensors like optical cameras, radar, lidar etc. An image processing unit combines the information form the different sensors and a decision making unit decides on the action the vehicle will take. Redundancy is required. If one of the (many) sub-systems fails, it is indispensable that redundant functions provide

© Springer-Verlag GmbH Germany, part of Springer Nature 2023
M. Hilgers, *Electrical Systems and Mechatronics*, Commercial Vehicle Technology,
https://doi.org/10.1007/978-3-662-66718-7_4

	Level 0	Level 1	Level 2	Level 3	Level 4	Level 5
	Manual driving – no automation	Assisted driving	Partial automated driving	Highly automated driving	Fully automated driving	Autonomous driving
Naming in SAE J3016	No Driving Automation	Driver Assistance	Partial Driving Automation	Conditional driving automation	High Driving Automation	Full Driving Automation
Driver's role	Driver is operating and monitoring the vehicle	Driver is operating and monitoring the vehicle	Driver permanently monitoring. No other task allowed	Driver is not driving. Driver in his seat to take over w. defined reaction time	Driver not in his seat. Takeover with long reaction time.	No driver
System's role	No "system"	Some function are assisted	System is performing longitudinal and lateral control .	System is driving. System might request the driver to take over with defined reaction time	System is driving. System is always capable of bringing vehicle to safe state.	System handles all situations independently
Example	Driver is driving. - Supporting functions like ABS/ESP not considered as assistance	Cruise control, adaptive cruise control	Traffic jam assistance drives the vehicle through traffic jam	System is driving under good conditions on the highway	On the highway the system is driving	Vehicle moves without outside intervention from start to destination

Fig. 4.1 Usually used classification of different levels to describe automated systems in on-road vehicles. Level 0 describes manual driving without any automation whereas level 5 describes the autonomous vehicle that does not require a driver anymore to complete the transport task from start to destination

enough information at least to bring the vehicle into a safe state ideally without obstructing traffic flow.

4.1   Environment Detection

For advanced driver assistance systems and for automated driving even more so, the control logic needs to obtain a picture of the environment around the vehicle. Exactly which details of the environment need to be recognized depends on the driver assistance system in question. From "level 2", partial automated driving, for example, the following objects must be recognized:

- The road and the lane must be recognized,
- other objects in the own lane must be detected
- objects on adjacent lanes or oncoming objects must be detected
- the speed of the objects must be determined
- road signs and traffic lights must be recognized
- etc.

With increasing automation, the detection of objects must be more accurate and additional objects must be detected. For example, an autonomous driving vehicle must recognize complex intersection situations or perceive the gestures of other road users and needs to understand them.

It is necessary to use different technologies in parallel for environment recognition to achieve the desired reliability. If one technology fails to detect an object for whatever reason the other technologies should recognize the object reliably. Optical cameras for example might be blinded by bright light or the low-standing sun. So currently the development activities on different technologies are trying to make high automated driving possible: Radar, lidar and camera systems to detect and observe the environment are used in parallel. The overall system for environment recognition must offer a sufficiently wide range, i.e. it must see a few hundred meters ahead to guarantee sufficient reaction time to the vehicle. The field of vision is also very important. The entire surroundings of the vehicle must be covered. And the overall system must achieve a high spatial resolution to distinguish different objects.

The different technologies to acquire information about the environment can be grouped into three classes:

First are passive systems, that use the "natural radiation" of objects. Objects reflect light from the environment or emit longer wavelength electromagnetic radiation. Light is defined as electromagnetic radiation that the human eye perceives without aids. This is approximately the wavelength range from 380[1] to 750 nm. Longer wavelength electromagnetic radiation is called infrared radiation. Sometimes ist is also called thermal radiation. Cameras in the optical range or infra-red cameras (so called thermal imaging cameras) perceive this radiation. There are mono cameras with only one lens and stereo cameras with two lenses generating two slightly displaced pictures that show the world from two (slightly) different angles. With this stereo-systems three-dimensional vision is possible—as in the human case with two eyes.

The human driver relies very much on the optical perception of the environment. In the optical range, additional information is transported by means of color. Color is the colloquial term for the spectral composition of light i.e., the frequency spectrum of electromagnetic radiation: Red and yellow are considered warning colors: signal lights, traffic signs or road markings are colored.

Passive infrared systems use infrared-sensitive camera to record the infrared radiation emitted by objects. This is the principle of a thermal imaging camera: Warmer objects (e.g., pedestrians or animals) emit more infrared radiation and deliver a clear signature.

Second, there are beam-based systems. This means that the vehicle emits a signal and measures the reflection of the signal. Other objects reflect the beam. Based on the time between emitting and sensing the pulse, distances to the objects can be determined. Using the Doppler-effect the speed of the object relative to the ego-vehicle is measured.

---

[1] The unit nm stands for nano meter $= 10-^9$ m.

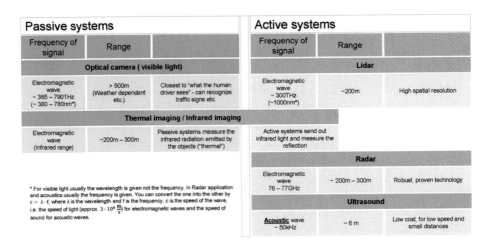

Fig. 4.2 Environment detection systems. Enabler for many ADA systems and automated driving. The technical data serves as orientation. Lidars, for example, are developed with different wavelengths and there are systems with different ranges. Generally, the longer the range, the smaller the angle covered by the system

Ego-vehicle describes the vehicle on which the assistance system is sitting. The beam-based system can be a radar system[2] that emits and detects electromagnetic radiation in the GHz range (very common are 76–77 GHz). Lidar systems[3] emit infra-red laser pulses and capture the laser light that is reflected. Infra-red means that the wavelength is outside the visible range (longer wavelength). Due to the short wavelength (compared to radar) of the laserlight, lidar offers high spatial resolution. Long range lidars with a range of more than 200 m and a wide field of vision are possible.

Figure 4.2 shows active and passive systems used in the automotive industry to sense the environment.

The third way of recognizing the environment is communication with other road users or the infrastructure. This is usually called vehicle-to-X communication, e.g., vehicle-to-infrastructure and vehicle-to-vehicle communication. This information channels allow to see behind the horizon: The vehicle might obtain information about road conditions or traffic long before it is visible for camera, lidar or radar. Also "invisible" information might be shared like low friction coefficients of the road or areas where the temperature is below zero in winter times. Moreover vehicle-to-X communication might allow to see into the future: a traffic light can communicate that it will turn to red in 15 s or other vehicles might communicate that they intend to change lanes or brake or take a

---

[2] Radar = radio detection and ranging.

[3] Lidar = laser imaging, detection, and ranging.

turn. For example, in intersection situations or when a vehicle has to make an emergency braking, vehicle-to-vehicle communication allows other road users to react earlier than would be possible using only the on-board data from the environment detection system. However, it must be considered that for the time being and for years to come not all road users will be able to participate in this kind of communication. Not all vehicles will have the technical capabilities to do so. Most probably there will always be a certain ratio of road users that do not participate in vehicle-to-X communication.

Environment recognition is a challenge for both sensor technology and image processing. The available imaging sensors are probably already sufficient for a very high level of automation. The challenge is to interpret the image sequences correctly, understand the current traffic situation and how it might evolve, and derive appropriate actions for the automated vehicle from that prediction. This process of observing, interpreting, predicting and acting must be repeated continuously and at a very high frequency. Road users react continuously to the other road users and the traffic situation changes continuously. The myriad of different traffic situations must somehow be handled in the control logic of automated systems.

4.2   Different ADAS Systems

There are some driver assistance systems that affect the longitudinal dynamics of the vehicle, and some that influence the lateral dynamics. Active intervention has already seen more progress in the area of longitudinal motion of the vehicle (i.e. braking and accelerating) than in lateral motion (i.e. steering). This is presumably because the road is longer than it is wide.

Another important area for ADAS are systems which aid the driver's visual perceptions. Figure 4.3 groups ADAS in categories.

Many driver assistance systems require relatively complicated technical equipment to be installed in the vehicle. But it is often the case that the electronic hardware already in the vehicle can be used for several systems. Figure 4.4 shows this with currently available driver assistance systems.

Some of the driver assistance systems described below have already been available for use in modern commercial vehicles for some time, others are recent developments or still in the development stage.

4.3   Advanced Driver Assistance Systems for Longitudinal Guidance

Cruise control

The cruise control function enables the driver to set a desired traveling speed. The vehicle travels at this speed without the driver having to continuously depress, or push, the

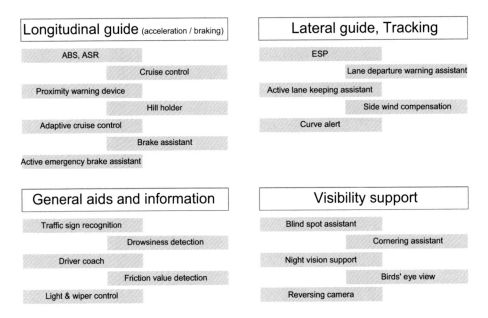

Fig. 4.3 Various advanced driving assistance systems (ADAS)

gas pedal. The driver can disable cruise control by operating the brake or with the control element for cruise control.

Collision warning system

The collision warning system uses radar to measure the distance from a car traveling ahead of the vehicle and warns the driver if the distance is or falls below a speed-dependent threshold.

Adaptive cruise control

Adaptive cruise control is a cruise control system that considers the traffic ahead. Without traffic in front the adaptive cruise control acts like a cruise control and moves the vehicle with the desired speed set by the driver. And it uses radar to detect vehicles ahead. The speed of the vehicles and the distance from the vehicle traveling ahead are measured. The adaptive cruise control maintains a speed-dependent safe distance to the vehicle ahead automatically. If the distance becomes smaller, the adaptive cruise control reduces the engine torque and slows the vehicle down. Adaptive cruise control performs partial braking actions; it applies the brakes in a manner that brings about a fraction (e.g. 30%) of the maximum possible deceleration. If the distance to the vehicle ahead falls below a threshold value and stronger braking is required, the adaptive cruise control emits an acoustic and visual alert to the driver and hands the vehicle control back to the driver.

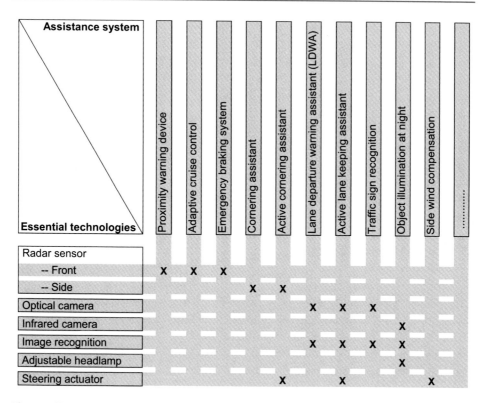

Fig. 4.4 Examples: various technical systems must be present in the vehicle to provide a number of different driver assistance systems. This list is not complete

If the driver presses the brake pedal or the accelerator the vehicle hands over the control to the driver, too. Some adaptive cruise control systems do not work below a certain speed: If the vehicle speed falls below a speed threshold because of the vehicle ahead the system as well hands over the control to the driver. Full range adaptive cruise control works down to standstill and might also start up again independently when the vehicle has come to a standstill and the vehicle ahead is starting up again. If the standstill period is longer than a defined timespan (e.g., 2 s), for safety reasons the vehicle does not start on its own but waits for a signal from the driver via the accelerator pedal or a steering wheel button.

Automatic emergency braking system (AEBS)

The automatic emergency braking system (AEBS) goes one step beyond the adaptive cruise control. If the distance from the vehicle ahead continues to get smaller despite the series of alerts and the driver has not responded appropriately, the emergency braking system initiates maximum full-stop braking automatically. The distance from the vehicle in front at which the maximum full-stop braking is carried out is calculated such that the vehicle comes to a standstill just in time under normal conditions. If the friction value between the road and the tire is lower than the value assumed in the emergency braking system algorithm, a collision may not be ruled out, but the collision speed is substantially lower. In this case, the emergency braking system has fulfilled the function of a collision mitigation system.

The AEBS has been available on the European truck market since 2006 [21, 22]. Emergency braking systems are subject to continuous further development, in particular with regard to the objects that the systems can detect (stationary objects, crossing pedestrians, etc.).

Friction value detection/Road condition detection

The AEBS might also be combined with friction value detection to define a minimum distance at which maximum full-stop braking is necessary. Detection of the friction value between the road and the tires also represents an additional aid for the driver independently of any emergency braking assistant. The vehicle might warn the driver if tire grip is low or if it changes suddenly. Research on friction value detection is ongoing [6]. Systems are discussed that combine all available information to output information on the road condition. The system can (like the human driver) use optical information from the cameras plus the information from the rain and light sensor. From the ESC-system individual wheel speed information, yaw sensor information and the like can be considered. The torque levels and the steering return torque are additional measurements that are discussed to determine road conditions [7]. Ambient temperature is an obvious value that is included into the system's logic. An infrastructure-backend can supply additional information. This backend might be fed by meteorological data, with data from the past and with data from vehicles ahead that used the road just a few minutes before [29]. However, systems relating to friction value detection are currently still in pre-development stages. Friction detection must provide reliable results before it can be offered in vehicles. If the system gives out unreasonable warnings or even triggers an unjustified braking, the driver will not accept the system and will most probably switch it off.

Brake assistant

With electronically controlled power brake systems, a brake assistant can be implemented easily [20]. The brake assistant electronically detects when the driver brakes hard and ensures that the full braking power is applied. Experience has shown that drivers do not immediately apply the brake with full force even in emergency situations; and consequently, precious stopping distance is wasted. The brake assistant makes up for this.

In order to determine whether the driver intends to carry out emergency braking, the electronics system evaluates the driver's behavior, particularly the speed with which the brake pedal is depressed as an indication that emergency braking is intended.

But unlike the AEBS, the brake assistant requires the driver to initiate braking and to unequivocally indicate his intention to brake hard.

Hill holder

The hill holder is another assistance system which uses the brake. It makes it easier to launch on hills. If the vehicle is sitting on a slope and the driver takes his foot off the brake pedal in order to drive away, the hill holder ensures that the vehicle does not roll backwards. For this, the brake is released gradually in synchronization with the buildup of torque on the rear axle so that the vehicle does not roll backwards.

4.4   Driver Assistance Systems for Lateral Guidance

Lane departure warning (LDW)

The heart of the lane departure warning[4] (LDW) is an optical camera located behind the windshield that scans the roadway in front of the vehicle. Image recognition software detects the lane markings. If there is a danger that the truck is going to wander out of its lane and the direction indicator has not been activated, the driver is alerted. An acoustic alert concept inside the truck emits a sound via the radio loudspeaker on the side to which the vehicle is wandering, so the driver can take corrective steering action intuitively. On coach buses, the driver is alerted by a haptic warning system: the steering wheel vibrates. The haptic warning can also be provided by vibration of the seat cushion on the side in question. In coaches, acoustic warnings to the driver are often avoided in order not to inconvenience the passengers.

The system prevents the driver from inadvertently driving out of the lane, thus helping to avoid dangerous traffic situations. It also creates a training effect. The driver becomes accustomed to driving in an orderly manner, keeping to the middle of the lane and activating the direction indicator when changing lanes.

However, very narrow lanes or unclear road markings can cause the system to issue an alert even though the driver is driving faultlessly. For this reason, the driver can switch the lane departure warning assistant off.

Frequent alerts to the driver from the lane departure warning assistant may also be a sign that the driver is either distracted or drowsy.

Active lane keeping assist

The LDW can be expanded with an active component. The vehicle cautiously adjusts the steering if it detects that it is about to depart from its lane. Creating an active lane keeping assist requires a steering actuator that enables the steering to be operated

---

[4] Or lane departure warning system (LDWS) or lane departure warning assistant (LDWA).

electronically. This may be implemented in the form of an electric motor on the steering shaft, for example.[5]

Instead of reacting to the vehicle approaching the lane marking/the lane borders the control logic of the active lane keeping assistant could define an ideal track (presumably very much in the middle of the lane) and responds to the deviations from this ideal track. Now if the deviation width permitted by the control is limited, the effect conceptually approaches that of a self-steering vehicle and autonomous driving.

Curve warning system
Conceivable is the curve warner, which knows the radius of the curve ahead based on the map of the navigation system. It compares the radius of the curve ahead with the current speed. If the speed is deemed to be too high the vehicle could warn the driver or even reduce the speed automatically. It should be noted that the friction value of the road, which changes rapidly and unpredictably, the height of the vehicle's center of gravity and the condition of the vehicle all play an important role in the evaluation of an appropriate cornering speed.

4.5   Advanced Driver Assistance Systems for Longitudinal and Lateral Control

The Traffic Jam Assist or Stop and Go Assist combines lateral and longitudinal control of the vehicle in one assistant system. It is a level 2 automated driving system—see Fig. 4.1. In traffic jam situations the vehicle independently follows the traffic ahead. It accelerates and brakes the vehicle, and it steers the vehicle to keep it in the lane. The driver does not have to perform the unnerving stop-and-go driving himself. He still has to monitor the system, though. If the system encounters a situation it can not handle (e.g. no proper lane markings, vehicle ahead is accelerating too quickly, etc.....) the system notifies the driver and the driver has to take over control of the vehicle.

As the stop and go assist is the first ADAS that combines lateral and longitudinal guidance of the vehicle it can be considered the entry into automated driving.

---

[5] But it is also conceivable, for the steering movement is produced by selective braking of individual wheel, similarly to an ESP control intervention. However, enabling the steering action by means of the brake also has drawbacks, for example, brake wear is increased and energy is lost (i.e. increased consumption). For these reasons, this approach is not recommended in the commercial vehicle segment.

4.6   General Aids Through Driver Assistance Systems

Traffic sign recognition

A driver assistance system that is found in many cars is traffic sign recognition. An optical camera scans the area close to the road. The image detection software detects certain traffic signs in the optical image, such as speed limits and no passing signs, and presents them in a suitable position in the display, possibly with a short warning tone to attract the driver's attention.

For higher levels of automated driving, it is essential to ensure reliable recognition of traffic signs, road markings, traffic lights and the like. The automated system will have to make decisions based on this. Vehicle to infrastructure communication can support the recognition of traffic guidance.

Driver attention detection (drowsiness detection)

Attention detection or drowsiness detection draws on various data supplied by the vehicle to deduce whether the driver is driving attentively or is at risk of falling asleep. The driver's movements of the steering wheel reveal characteristic patterns when the driver becomes less attentive. If a camera monitoring the street and the lane markings is present, the information from the camera for the lane keeping assistant (LKA) can also be evaluated. Based on additional information such as operation of direction indicators, gear shifting, etc., it is also possible to deduce whether the present driving situation is rather monotonous or more varied. If the driver is assessed as being inattentive, a warning signal is emitted and/or a message appears in the display suggesting that the driver takes a break.

There are also solutions with a camera for observing the driver's face and diagnosing insufficient attentiveness with the aid of image recognition software. Such a system coupled with a proximity warning system can detect that the driver is not watching the road while approaching an obstruction, for example. Other concepts which are also in pre-development capture signs of drowsiness on the part of the driver with a camera directly, for example, by analyzing blinking patterns or eye movements.

In countries where privacy protection is less strict the information from the camera observing the driver can also be used to monitor other driver behaviour like smoking in the truck or looking at their mobile phone. If the system detects any behavior that is considered misbehavior the driver can be warned or a warning (with the respective video capture) might even be sent to the freight forwarding company. This then is obviously not a driver assistance anymore.

Light and windshield wiper control

There are also intelligent light controls to help the driver feel safe and relaxed. These assistance systems are described in the section on lighting, see Sect. 1.3.2.

The windshield wiper controller is also a driver assistance system in the broadest sense. A rain sensor detects water on the windshield and actuates the windshield wipers.

Tire pressure monitor

The tire pressure monitor informs the driver about the current tire pressure and also about the desired target pressure [3]. If the difference between target pressure and actual measured pressure is greater than a certain threshold value, the driver is warned.

Driver coach

The driver coach evaluates the driver's driving style during the journey and provides the driver with direct feedback, possibly even tips on how to improve driving behavior. There are onboard driving coaches, which interact directly with the driver while driving, and there are telematics systems that perform driving style evaluation offline after the data has been transmitted telematically from the vehicle. The online driving coach must be precise and concise in order to help the driver without distracting him too much from the driving task.

The primary purpose of driving style training is most likely to optimize fuel consumption. In this context, the focus may be on how erratically the driver operates the gas pedal, whether the driver accelerates aggressively, how often and how sharply brakes are activated, or whether the wear-free brake (i.e. retarder) is utilized. Driver coaches designed to promote economical driving habits have been available for installation in serial production vehicles since 2011.

Not currently available but technically just as feasible are driving style trainers that also evaluate safety aspects: If the vehicle is fitted with traffic sign recognition, an assessment might be made as to how conscientiously the driver obeys speed limits. Vehicles equipped with a radar eye might assess the distance between the Ego-vehicle and the vehicle ahead. And if the vehicle is equipped with lane detection and a sufficiently high-resolution traffic sign recognition device, the driver could be warned against overtaking in a no passing zone. The above-mentioned curve warning system could be considered a part of a safety-centric driving style trainer, too.

Remote controlled maneuvering

Trucks with advanced driver assistance systems have to have an electronical interface to control the longitudinal (acceleration, braking) and lateral (steering) movement of the vehicle. This interface can be used to control the vehicle remotely via a radio connection.

One variety for remote controlled maouvering is that the driver is controlling the vehicle but standing outside the vehicle. The remote control unit might be a separate device or it could be an App on a mobile device. If the driver is controlling the vehicle remotely no electronic brain to determine the driving is required. The driver is in control.

A second variety is that e.g. on larger haulage yards there is a local infrastructure with cameras and other sensors and an advanced control logic (outside the truck) that is controlling the yard and (all) the vehicles on it. This system determines where the truck should go and remotely controls the truck accordingly.

Other functions that are needed for loading and unloading like the lighting system, the chassis level control system or PTOs might be included in the control logic as well.

In any case the interface to operate the truck remotely has to fulfill very strict security standards.

4.7   Driver Assistance Systems in Support of Visibility

Blind spot assistant, Blind Spot Detection and Cornering assistant
The blind spot assistant (BSA) uses radar sensors to check whether other road users are in the driver's blind spot when changing lanes. When the driver activates the turn signal to indicate the intention to change lanes, the BSA emits an acoustic and/or optical warning signal if another road user is in the hazard area.

Other systems focus on cornering in urban traffic. When the vehicle is stopped at an intersection, the cornering assistant checks the area in front of the vehicle and on the passenger side of the vehicle. If another road user or an obstruction is detected in this hard-to-see area, the system warns the driver of this.

If the vehicle is equipped with active steering, (for example, electric power-assisted steering), it is possible that the BSA may also provide a haptic warning with a movement of the steering wheel not to change lanes in addition to the acoustic and optical warnings.

Reversing camera
The reversing camera makes reversing easier for the driver by showing an image of the area behind the vehicle that the driver otherwise cannot see. The camera image appears on a display in the cockpit.

Bird's Eye View
Bird's Eye View is a driver assistance system that provides the driver with a picture of their vehicle and the area around the vehicle from an aerial perspective. A number of wide-angle cameras are arranged at various points on the vehicle. The pictures from these cameras are merged in a computer to produce a synthesized overall image that shows an artificial aerial view of the vehicle and the area around it. This image appears on a display in the cockpit. In order to generate a complete picture of the entire vehicle, cameras must be located on all sides of the vehicle. Since tractor semitrailer combinations and articulated truck trailer combinations are designed to change trailers frequently and since semitrailers and trailers have a bend in the vehicle combination when cornering it is easier to produce bird's eye view images for rigid trucks. The bird's eye view function may be particularly helpful for vehicles that are operated in applications where people are working constantly around the truck. Garbage trucks are suitable use cases, for example.

Night vision system

Another possible driver assistance system is the night vision system. This consists of infrared headlamps and a camera that operates in the infrared range. The street is illuminated with infrared lighting and observed with the camera. The simplest night vision systems show the driver the images from the infrared camera on a screen in the vehicle. However, the disadvantage of this system is that the driver must watch the road and the night vision system screen at the same time. The next level of night vision assistance couples the system with image detection function. Then, the driver's attention only needs to be drawn to the screen by a distinctive acoustic and/or color signal when objects of concern have been detected (pedestrians, animals, etc.). It might also be possible to combine this with a heads-up display so that the driver does not constantly have to switch his attention back and forth between the road and the screen.

Another approach is to couple the night image detection system to a movable headlamp that shines on objects of interest identified by the image detection system and renders them visible to the driver. The advantage of this approach is that the driver's attention is not distracted from the road and there is no need for a screen or heads-up display that the driver has to watch.

Highly and fully automated vehicles can make good use of infrared based additional vision information. The information can give valuable input to capture the surrondings of the vehicle and take decisions.

4.8   Autonomous Driving

The assistance and telematic systems that are already available for the market point into the direction of the ultimately autonomously driving truck. Distance-maintaining cruise control systems, stop-and-go assistant or traffic jam assistant and emergency braking assistants already control the longitudinal guidance of the vehicle in many situations without human intervention. Combined with a high-performance traffic sign detection system, it is technically not difficult to adapt the speed settings automatically to the speed limits in a given situation. Even now, traffic moving ahead of the vehicle and the speed at which it is traveling is already detected reliably with radar systems and an optical camera.

An active lane keeping assistant can be restricted so that the vehicle follows a predetermined track. The truck is guided in the middle of its lane fully automatically by steering interventions. However, the roadway detection capability in an autonomous steering system must satisfy requirements that are far more stringent than those that apply for assistance systems.

The first autonomous driving truck was approved for driving on a public highway in 2015 [24]. [23] shows how far technical development has advanced: On a defined section of road, an autonomous driving truck was moved in the flow of traffic with other vehicles without driver intervention. Technologically, the vehicle was equipped with a radar

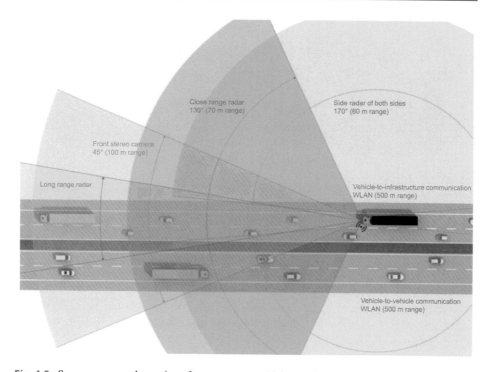

Fig. 4.5 Sensors on an early version of an autonomous driving truck. (Illustration: Daimler)

sensor for scanning long and short ranges forward, a forward-looking stereo camera and other radar sensors that were responsible for monitoring the roadway to the right and left of the truck—see Fig. 4.5. Future systems for autonomously driving trucks probably will have lidars as well (see section 4.1). The information from the sensors is interlinked (sensor fusion) and provides a complete picture of the surroundings in the central computer. Digital map data can provide still more information.

It is to be expected that autonomous driving vehicles will receive more pertinent information through vehicle-to-vehicle communication (V2V) and vehicle-to-infrastructure communication (V2I).

With V2V communication, acceleration and braking maneuvers of vehicles ahead of the autonomous vehicle can be captured even when these vehicles are not yet within the detection range of the camera or radar. The vehicle can calculate with information that sees even further ahead. Braking actions by traffic traveling ahead (i.e. any vehicle not yet visible) due to fog banks, accident sites or traffic congestion, for example, can thus be detected and the vehicle approaching these obstacles can adjust its speed in good time. Besides the gain in terms of safety, this also offers fuel consumption benefits. With V2I communication, it is conceivable that additional information not included in the digital map can be transmitted to the vehicle. This may include, for example, information

Fig. 4.6 Left: Prototype for autonomous driving based on an established long-haul tractor vehicle with comfortable cab. Right: Prototype for a tipper truck without a cab. As there is no cab, the dimensions of the autonomous vehicle are difficult to judge: this is a tipper with a 15 t haulage capacity. (Photos: Left: Daimler, right: Volvo Construction Equipment)

about one-day or moving roadworks. Changeable traffic signs (i.e. speed limits, altered lane markings) can also be transmitted to the vehicle. Conversely, the information from the vehicles might be used by the infrastructure. Traffic congestion and inconsistent vehicle speeds might be sent to the infrastructure and processed in traffic control systems or traffic information reports.

An autonomously driving truck must be able to cope with a myriad of different traffic situations. Today's systems are not yet there. Research is ongoing and the number of companies developing fully automated and autonomous trucks or at least contributing to this development is huge. The companies interested in developing systems for autonomous driving range from the big tech giants, includes vehicle OEMs and suppliers and goes down to small start-ups.

The prototypes that are shown are mostly based on existing truck models or copy the well-known truck concept with a traditional cab—see Fig. 4.6 on the left. This is—at least in the development stage—a very understandable approach as there might always be a situation where the system fails, and a human driver must intervene. Although in a really autonomous vehicle the cab could be omitted, reducing cost and giving additional design options. In gated areas, that are permanently monitored by a central control room, trucks without cab seem feasible. These areas could be mining areas, container terminals at large harbors, large construction sites, industrial production sites, big agricultural units or the like. Figure 4.6 on the right shows a prototype for a tipper that does not have a cab anymore. In any case gated areas or predefined routes are a good starting point for autonomous driving vehicles: the traffic situations the system will encounter are less diverse. Favourable boundary conditions can be created like clear lane markings or additional visual or electronic landmarks that help the autonomous system to find its way. Certain areas might be restricted for other road users, in particular for pedestrians. Speed limits might also help.

Many of the technologies that will be used for the "autonomous driving" function will be used for commercial vehicles and passenger cars alike. But of course there are challenges specific to commercial vehicles, such as the larger dimensions and the trailer running along a trajectory that deviates from that of the towing vehicle in curves.

The speed at which autonomous driving will be able to establish itself will also depend on whether and how quickly a suitable legal framework can be found.

Telematic Systems

5

Systems that exchange data between the vehicle and some data backend outside the vehicle are often discussed using the umbrella term telematic systems. Telematics is a newly coined word from the two terms telecommunications and informatics. The information exchange used for automated driving (vehicle-to -vehicle and vehicle-to -infrastructure) is usually not subsumed under telematics. Telematics is usually used to describe data exchange that supports the business side of the freight forwarding company. The various services of telematic applications for commercial vehicles can be organized into three groups [17]: transport management, vehicle management and time management. There are a number of different providers of telematic systems for commercial vehicle goods transport.

Transport management embraces various telematic services that make it easier for the fleet and the driver to deliver the goods to their destination on time. They include navigation, but also route monitoring by the fleet. With telematics, the fleet manager can be kept informed about the location of his vehicles at any given time. It is then possible to see in real time if the vehicles are completing their routes as planned or if there is a need to change the route planning. The transmission of refrigeration data or the like is also possible with telematic services. Various telematic systems also integrate the handling of package items in the telematic systems directly. With a barcode scanner and signature capture, the entire goods handling can be mapped in the telematics system.

The vehicle management part of telematic systems concerns itself with the vehicle and how the driver works with it. Data about the condition of the vehicle is transmitted to the fleet. They can keep track of fuel consumption by each individual vehicle or monitor vehicle speeds. Detailed consumption data and a route dependent usage analysis, that provides a measure of how demanding a route is, make it possible to analyze whether the driver is driving economically.

The time management services of the telematic solutions allow the data from the digital tachograph to be read out from the vehicle via mobile data connection and analyzed

© Springer-Verlag GmbH Germany, part of Springer Nature 2023

M. Hilgers, *Electrical Systems and Mechatronics*, Commercial Vehicle Technology,

https://doi.org/10.1007/978-3-662-66718-7_5

Table 5.1 Examples of various speed assumptions in the navigation algorithm of truck navigation systems

Road	Postulated average speed
Freeway	70 km/h (44 mph)
Open road	55 km/h (34 mph)
Urban traffic	30 km/h (19 mph)

and stored by the fleet. In the offices (offline), this data is then used to plan the driver's next work assignments, and the data may possibly be used for calculating payroll as well.

Navigation

Probably the most important telematic system from the driver's point of view is the navigation system. Navigation systems are already in widespread use and can be found not only in automobiles but also on mobile phones, for example.

The quality of the route guidance offered by the navigation system is determined by three major factors: the algorithm used by the system, the accuracy and freshness of the digital map data, and the quality of the traffic information with which the navigation system works.

Truck navigation is characterized by some special considerations and differences from car systems. The most important of these are roads where commercial vehicles are not permitted or for which restrictions apply, which make certain routes unusable for trucks (height, vehicle weight, axle load, ferries that are only suitable for cars, etc.).

The algorithm with which the optimum route is worked out must assume certain average speeds for different roads in order to be able to compare different routes (Table 5.1). These premises may differ from one navigation system to another, so different navigation systems may suggest different routes or predict different driving times for the same route.

Vehicle locating and geofencing

Telematic systems allow the fleet to find out in real time where a given vehicle is. This makes scheduling the vehicles easier. The scheduling department can track how well the vehicles are proceeding with their respective routes online. The dispatcher can see which vehicle with available cargo space is currently close to a specific loading point, so that it may be able to take on an extra transport assignment at short notice.

The location information can also be used to protect the vehicle, when the vehicle is only authorized to operate in a certain geographical territory. The vehicle cannot leave this territory without authorization. This might be helpful for truck rentals and protects the vehicle from being stolen. With a localization device stolen vehicles can be located again easily.

It is also conceivable to install locks on the cargo area doors that can only be opened at certain locations or must be released with a radio signal from headquarters when the vehicle has reached the unloading area, to prevent unauthorized unloading.

Driving style evaluation

Driving style analysis can be carried out onboard with a driving coach installed in the vehicle (see Sect. 4.3 driver assistance systems). Alternatively, or additionally detailed information about the route and the driving behavior on the route can be transmitted from the vehicle to a server via a telematic system. This way, the driver's driving style can be evaluated subsequently, and economical practices can be recognized. The fuel consumption of drivers on different routes can be made comparable by taking their route profile into account in the evaluation. Freight forwarders use this capability to make driving performance of drivers comparable and they may use this as a basis for paying individual bonuses to good drivers. In turn drivers become more actively committed to improving their driving style. Intercompany competitions encourage the drivers to drive in a way that is evaluated favorably in terms of fuel consumption and wear.

Other data that can be evaluated by the freight forwarder relates, for example, the question as to whether the driver is wearing their seatbelt, or the driver assistance systems are switched off.

Transfer of information to the vehicle

Of course, the communication path from freight forwarder to the vehicle can also be used to provide the driver with information such as new routes, additional unloading or loading locations, less congested roads, or even birthday greetings. In cases of absolute necessity, the driver can be guided in detail with step-by-step work instructions.

Automatic emergency call

A telematic system designed to mitigate the consequences of an accident is the automatic emergency call. If the vehicle senses a serious accident, an emergency call is automatically transmitted to a universal emergency call number. The idea behind this is to enable aid and rescue services to get to the scene more quickly and by doing so improve the victims' chances of survival and recovery. The automatic emergency call transmits the location of the accident and the direction in which the vehicle was traveling to the emergency call center. The vehicle's travel direction is particularly important for aid services on roads with divided carriageways, (i.e. freeways, interstate highways, etc.).

It is also conceivable for the emergency call function to establish a voice link between the call center and the vehicle involved in the accident. If the occupants of the damaged vehicle are able to talk, they can also supply further information.

The automatic emergency call function also has provision for manual activation, so that witnesses to an accident who are not directly involved can also report the accident from their own vehicle.

In order to enable an automatic emergency call to be made, technical equipment must be specified on the vehicle. This particularly includes a communication unit and a unit that recognizes a serious accident as such. In a car, an accident can easily be detected via crash sensors, which must be installed for the airbag. Nowadays, most trucks are not equipped with crash sensors, so additional sensors have to be specified in the truck.

An emergency system, named eCall, is currently mandatory for light vehicles in Europe.

Toll device

In some countries, charges for road use (i.e. tolls) are collected. In order to make this process easier for truck drivers, and complete their drive with as little disruption as possible, there are often devices installed in the vehicle that transmit information to the toll system so that the system can calculate the toll charge payable and the payment can made automatically.[1]

The device has a positioning function (via GPS) and knows various vehicle characteristics such as its number plate, registration number, permissible gross weight, number of axles, hazardous materials classification, and other details that are used to identify the vehicle and determine the toll to be paid. The device includes a wireless interface via which it can exchange information with the toll system infrastructure.

The tolls systems differ from one country to another, and the devices are also not standardized. Different countries require different devices. It is not unknown for a vehicle to have several devices installed to meet the different requirements of several different countries.

---

[1] In all countries there are also ways to pay tolls if the vehicle is not specified with such a device. It is not hard to imagine a situation in which a given truck only travels once through the country in question, so it is not worthwhile to install the device.

Comprehension Questions

The comprehension questions serve to test how much the reader has learned. The answers to the questions can be found in the section to which the respective question refers. If it is difficult to answer the questions, it is recommended that you read the relevant sections again.

A.1 Mechatronics
What are the components of a mechatronic system?

A.2 Onboard Electrical System 1
a) What technology is used in standard automotive starter batteries?
b) What is the nominal voltage of the onboard electrical system in a heavy commercial vehicle?
c) Why does a truck have two batteries? How are they connected?
d) How is the battery charged?

A.3 Onboard Electrical System 2
a) How is the energy content of the battery indicated?
b) What are typical energy contents for a truck battery?
c) Why does a vehicle consume no-load current even when it is switched off?
d) How long does it take to drain the truck batteries?

A.4 Sensors
Explain the measurement principle of a:

a) Temperature sensor,
b) Pressure sensor,
c) Yaw rate sensor,
d) Condensation sensor.

© Springer-Verlag GmbH Germany, part of Springer Nature 2023
M. Hilgers, *Electrical Systems and Mechatronics,* Commercial Vehicle Technology,
https://doi.org/10.1007/978-3-662-66718-7

A.5 Bus System

a) What is CAN bus?

b) What bus topologies are there?

A.6 Compressed Air System

a) Why are most trucks equipped with compressed air brakes?

b) What components does the vehicle have for delivering compressed air?

c) What functions—besides the brakes—use the compressed air?

A.7 Wheel Brake

What kinds of wheel brakes are there?

A.8 Brake Cylinder

Explain the combination brake cylinder.

A.9 Brake Functions

Explain the following functions:

a) ABS function,

b) ASR function,

c) ESP function.

A.10 Brake Functions

Explain the following terms: auxiliary braking system and power-assisted brake system (these two terms may sound similar, but they refer to two very different things).

A.11 Automated Driving

a) What does "level 2" automated driving mean?

b) How many levels are there to classify different levels of automation?

A.12 Environment Recognition

Explain the different systems used to perform environment recognition for automated driving. What are the advantages and disadvantages of the different technologies?

Abbreviations and Symbols

The following is a list of the abbreviations used in this booklet series. The letters assigned to the physical variables are in conformity with normal usage in the engineering and natural sciences.

The same letter can have different meanings depending on the context. For example, lower case c is a busy letter. Some abbreviations and symbols have been subscripted to avoid confusion and improve the readability of formulas, etc.

Lowercase Latin Letters

a	acceleration
bar	bar, unit of pressure measurement—1 bar $= 10^5$ Pa
c	coefficient, constant of proportionality
c	usually used for the speed of light, approx. $3 \cdot 10^8$ m/s
da	abbreviation for deca $= 10$, used particularly often to indicating the measurement of force daN (deca-Newton), because 1 daN $= 10$ N is approximately equal to the gravitational force of one kilogram on Earth
f	correction factor coefficient
f	frequency
g	gravitational accelaration $= 9.81$ m/s^2
g	gram, unit of mass
h	length (often height)
k	kilo $= 10^3 =$ a thousandfold
kg	kilogram, unit of mass
km/h	kilometers per hour—unit of speed; 100 km/h $= 27.78$ m/s
kW	kilowatt—unit of power; one thousand watts
kWh	kilowatt hour, unit of energy
l	length

© Springer-Verlag GmbH Germany, part of Springer Nature 2023
M. Hilgers, *Electrical Systems and Mechatronics,* Commercial Vehicle Technology,
https://doi.org/10.1007/978-3-662-66718-7

l	liter, volume measurement; $1\,l = 10^{-3}\,m^3$
m	mass
m	meter, unit of length
m	milli $= 10^{-3} =$ one thousandth
mol	mole, unit of amount of substance—$1\,mol = 6.022 \cdot 10^{23}$ particles
n	number of particles, amount of substance
n	nano $= 10^{-9}$
p	pressure
r	radius (linear measurement)
rpm	revolutions per minute, unit: $min^{(-1)}$—used as a synonym for rotational speed as well
s	distance (linear measurement)
t	time
t	ton—unit of mass; $1\,t = 1000\,kg$
v	speed (velocity)
z	braking ratio, ratio between the vehicle's brake force and its weight force

Uppercase Latin Letters

A	Ampere, unit of electric current
A	area, particularly face area
ABA	active brake assist—the term used at Daimler for AEBS
ABS	anti-lock braking system
AEBS	autonomous emergency braking system
Ah	amp hour, unit of electrical charge—see C, Coulomb. $1\,Ah = 3600\,C$
ASR	anti-slip regulation from German Anti-Schlupfregelung—same as TCS
BSD	blind spot detection
C	Celsius (centigrade), unit of temperature
C	Coulomb, unit of electrical charge
C*	brake factor
CAN FD	controller area network with flexible data-rate, bus technology
CAN	controller area network, bus technology
CO_2	carbon dioxide
ADAS	advanced driver assistance system
DIN	Deutsches Institut für Normung (German institute for standardization)
DSR	differential slip control
E	energy

ECU	electronic control unit	
EDP	electronic data processing	
EMC	electromagnetic compatibility	
EPB	electropneumatic brake	
ESC	electronic stability control	
ESP	electronic stability program	
ETC	European transient cycle—Test procedure for emissions legislation	
F	force	
F_G	gravitational force	
G	Giga $= 10^9 =$ Billion	
GPS	global positioning system	
HiL	hardware in the Loop, test setup with which the function of (multiple) electrical control unit(s) is tested	
HMI	human–machine interface	
Hz	Hertz, unit of frequency, 1 Hz $= 1$ s^{-1}	
J	Joule, unit of energy	
K	Kelvin, unit of temperature on the Kelvin scale	
LDWS	lane departure warning system	
LED	light emitting diode—see Sect. 1.3	
LIN	local interconnect network, bus technology	
M	Mega $= 10^6 =$ Million	
M	torque	
MJ	MegaJoule, unit of energy—one million Joules	
MMI	man–machine interface	
MW	MegaWatt, unit of power—one million Watts	
N	Newton, unit of force	
N_A	Avogadro constant	
NTC	negative temperature coefficient thermistors	
OEM	original equipment manufacturer	
OTA	over the air—means via a radio connection	
P	power	
PSI 5	peripheral sensor interface 5, bus technology	
RTE	runtime environment, software interface in the Autosar standard	
SiL	software in the Loop, test procedure in which software (which will subsequently run in an electrical control unit) is tested for functionality on a test computer	
SOC	state of charge, measurement of a battery's charge level	
SOH	state of health measurement of a battery's aging status	
SWC	software component	

T	temperature (in Kelvin or °C)
T	Terra $= 10^{12}$
TCO	total cost of ownership; total costs incurred over the useful life of the vehicle or other
TCS	asset
	traction control system
V	volt, unit of electric potential
V	volume
V2I	vehicle to infrastructure communication
V2V	vehicle to vehicle communication
W	mechanical work or mechanical energy
W	watt, unit of power
Wh	watt hour, unit of energy—see also the more commonly used kWh
W_{kin}	kinetic energy

Lowercase Greek Letters

α *(alpha)*	angle
β *(beta)*	angle
γ *(gamma)*	angle
λ *(lambda)*	wavelength
μ *(mu)*	coefficient of friction, sometimes also μ_k
μ	stands for micro $= 10^{-6} =$ millionth
μC	abbreviation for microcontroller
ϕ *(phi)*	angle

References

General Reference Works

1. Wallentowitz, H., Reif, K. (eds.): Handbuch Kraftfahrzeugelektronik. ATZ/MTZ Specialist publication. Vieweg, Wiesbaden (2006)
2. Trautmann, T.: Grundlagen der Fahrzeugmechatronik. ATZ/MTZ Specialist publication. Vieweg Teubner, Wiesbaden (2009)
3. Hilgers, Michael: Chassis and Axles. 2nd edition. Commercial Vehicle Technology. Springer, Berlin (2023)
4. Hilgers, Michael: Alternative Powertrains and Extensions to the Conventional Powertrain. 2nd edition. Commercial Vehicle Technology. Springer, Berlin (2023)

Technical Articles

5. DIN 72552, Klemmenbezeichnungen in Kraftfahrzeugen; sheets 1 to 4
6. Bian, N., et al.: Fusion von Fahrzeug und Umgebungssensorik. ATZ **112**(9), 614–620 (2010)
7. Degerman, P., Anund, O.A.: Friction estimation using self-aligning torque for heavy trucks. Chassis. tech, 2nd International Munich Chassis Symposium, Munich, Germany, June 7 and 8 (2011)
8. ECE-R 10 Agreement concerning the adoption of uniform technical prescriptions for wheeled vehicles, equipment and parts which can be fitted to and/or be used on wheeled vehicles and … Regulation No. 10 Uniform provisions concerning the approval of vehicles with regard to electromagnetic compatibility
9. http://www.autosar.org
10. Robert BOSCH GmbH, Stuttgart: CAN Specification version 2.0 (1991)
11. Robert BOSCH GmbH, Stuttgart: CAN with Flexible Data-Rate, Specification Version 1.0 (released April 17th, 2012)
12. Robert BOSCH GmbH, Stuttgart: CAN with Flexible Data-Rate, White Paper, Version 1.1. (2011)
13. LIN Consortium: LIN Specification Package, Revision 2.0. Older and more recent specifications are available. http://www.lin-subbus.org (2003). Accessed July 2012
14. PSI5: Peripheral Sensor Interface for Automotive Applications. Technical Specification V 2.0. http://www.psi5.org (2011). Accessed July 2012

© Springer-Verlag GmbH Germany, part of Springer Nature 2023
M. Hilgers, *Electrical Systems and Mechatronics,* Commercial Vehicle Technology,
https://doi.org/10.1007/978-3-662-66718-7

15. NXP Semiconductors: UM10204, I2 C-bus specification and user manual (2014)
16. Daimler AG, Stuttgart: Actros 963 operator's manual (2011)
17. Daimler FleetBoard GmbH. http://www.fleetboard.com (2012). Accessed 17 June 2012
18. ECE-R 13. Regulation No. 13 of the UN Economic Commission for Europe (UN/ECE)—Uniform provisions concerning the approval of category M, N and O vehicles with regard to brakes
19. Pressel, J., Reiner, M.: Die Basisregelstrategie der elektropneumatischen Bremsanlage (EPB) von Mercedes-Benz. In: Reifen, Fahrwerk, Fahrbahn VDI reports, Vol. 1224 (1995)
20. Pressel, J., Reiner, M.: Weiterentwicklung des Telligent-Bremssystems im Actros. In: VDI-Gesellschaft Fahrzeug und Verkehrstechnik: Nutzfahrzeuge mit tragenden Lösungen ins nächste Jahrtausend VDI reports, Vol. 1504 (1999)
21. Scherhaufer, I., et al.: Active Brake Assist—Erfahrungen aus 4 Jahren Serieneinsatz. DEKRA VDI-Symposium 2010, Safety and commercial vehicles, Germany, October 28 and 29, 2010 (2010)
22. Scherhaufer, I., et al.: Active Brake Assist—6 Jahre Serienerfahrung. Chassis. tech, 3rd International Munich Chassis Symposium, Munich, Germany, June 21 and 22 (2012)
23. Daimler: Press release: Mercedes-Benz Future Truck 2025: Autonomes Fahren im Lkw-Fernverkehr mit dem "Highway Pilot", July 3 (2014)
24. Daimler: Press release: Freightliner Inspiration Truck—Der erste autonom fahrende Lkw mit US-Straßenzulassung, May 5 (2015)
25. FMVSS FEDERAL MOTOR VEHICLE SAFETY STANDARDS—Standard No. 121; Air brake systems
26. SAE J3016: Taxonomy and Definitions for Terms Related to Driving Automation Systems for On-Road Motor Vehicles. Revised 2021-04
27. Wissenschaftlicher Dienst des Deutschen Bundestages: Autonomes und automatisiertes Fahren auf der Straße—rechtlicher Rahmen, WD 7-3000-111/18 (2018)
28. Altherr, A.: Vision einer autonomen Landwirtschaft. CVC Jahrestagung 22nd Nov 2019 (2019)
29. Lauxmann R. et al.: Von assistierten Sicherheitsfunktionen zu … . ATZ, 120 Jahre ATZ (2018)

Index

© Springer-Verlag GmbH Germany, part of Springer Nature 2023
M. Hilgers, *Electrical Systems and Mechatronics,* Commercial Vehicle Technology,
https://doi.org/10.1007/978-3-662-66718-7

Printed in the United States
by Baker & Taylor Publisher Services